MECHANISMS OF OXIDATION BY METAL IONS

REACTION MECHANISMS IN ORGANIC CHEMISTRY

A SERIES OF MONOGRAPHS

MONOGRAPH 10

MECHANISMS OF OXIDATION BY METAL IONS

D. BENSON, B.Sc., Ph.D., F.R.I.C.

Halton College of Further Education,
Widnes, Cheshire, Great Britain

ELSEVIER SCIENTIFIC PUBLISHING COMPANY
AMSTERDAM - OXFORD - NEW YORK   1976

ELSEVIER SCIENTIFIC PUBLISHING COMPANY
335 JAN VAN GALENSTRAAT
P.O. BOX 211, AMSTERDAM, THE NETHERLANDS

AMERICAN ELSEVIER PUBLISHING COMPANY, INC.
52 VANDERBILT AVENUE
NEW YORK, NEW YORK 10017

ISBN: 0-444-41325-1

WITH 20 ILLUSTRATIONS

PRINTED IN THE NETHERLANDS

To my wife and my children

PREFACE

Metal ions are widely employed as oxidants in synthetic organic chemistry. Experience has shown that their reactivity can be controlled, in the main, by the correct choice of oxidant and of reaction conditions and, stemming from the pioneering work of Westheimer and the later studies made by Waters and his group at Oxford, there has been much interest shown in the kinetic and mechanistic features of their oxidation of organic compounds. I have tried in this little book to provide a concise but fully documented account of this developing subject. However, since I feel that it is unwise to consider organic oxidations in isolation, I have summarised, in addition, the main aspects of the oxidation of inorganic species.

I am indebted to Dr. L.H. Sutcliffe for advice and to a number of authors for allowing me to make use of diagrams from their publications. Mrs. E. Geraghty and Mrs. F. Martin were my helpful and efficient typists.

D. Benson

CONTENTS

Chapter 1

INTRODUCTION

In terms of their role in elementary kinetic steps, metal ion oxidants function either as one-equivalent or as two-equivalent reagents. One-equivalent oxidants are those which accept a single electron by direct transfer or by interaction with a hydrogen atom. Two-equivalent oxidants accept two electrons from the substrate; alternatively, they accept a hydride ion or lose an oxygen atom. These possibilities are illustrated in the case of permanganate by the following series of reactions.

$$Mn^{VII}O_4^- + MnO_4^{2-} \longrightarrow Mn^{VI}O_4^{2-} + MnO_4^- \quad \text{one-electron transfer}$$

$$Mn^{VII}O_4^- + HCOO^- \longrightarrow HMn^{VI}O_4^- + \cdot CO_2^- \left.\rule{0pt}{24pt}\right\} \text{hydrogen atom transfers}$$
$$Mn^{VII}O_4^- + R_2CHO^- \longrightarrow HMn^{VI}O_4^- + R_2\dot{C}-O^-$$

$$Mn^{VII}O_4^- + HCOO^- \longrightarrow Mn^{V}O_4^{3-} + CO_2 + H^+ \quad \text{two-electron transfer}$$

$$Mn^{VII}O_4^- + HCOO^- \longrightarrow HMn^{V}O_4^{2-} + CO_2 \left.\rule{0pt}{24pt}\right\} \text{hydride ion transfers}$$
$$Mn^{VII}O_4^- + R_2CHO^- \longrightarrow HMn^{V}O_4^{2-} + R_2C=O$$

$$Mn^{VII}O_4^- + CN^- \longrightarrow Mn^{V}O_3^- + OCN^- \quad \text{oxygen atom transfer}$$

Transfer of a halogen atom is also a possible route. Clearly, transfer of protons, hydroxide ions or oxide ions does not constitute an oxidation process since the oxidation state of the oxidant remains unchanged, e.g.

$$Mn^{VII}O_4^- + H^+ \longrightarrow HMn^{VII}O_4$$

$$Mn^{VII}O_4^- \longrightarrow Mn^{VII}O_3^+ + O^{2-}$$

$$Mn^{V}O_4^{3-} \longrightarrow Mn^{V}O_3^- + O^{2-}$$

Reactions involving a multi-equivalent oxidation as a single stage are considered unlikely. Indeed, even single-stage, two-equivalent processes seem comparatively rare.

Chapter 2, the longest part of the book, deals with reactions of a group of one-equivalent metal ion oxidants, the most important of which are cobalt(III), cerium(IV), vanadium(V), manganese(III), iron(III), and copper(II). In their reactions with organic substrates, the most frequently encountered oxidation process would seem to correspond to an electron transfer between substrate and oxidant accompanied by the breaking of a C—H bond and loss of a proton to leave a substrate radical, e.g.

$$Co(III) + C_6H_5CH_3 \longrightarrow Co(II) + C_6H_5CH_2^{\cdot} + H^+$$

$$Co(III) + C_6H_5CH_2OCH_2C_6H_5 \longrightarrow Co(II) + C_6H_5\overset{\cdot}{C}HOCH_2C_6H_5 + H^+$$

$$Mn(III) + (CH_3)_2CHOH \longrightarrow Mn(II) + (CH_3)_2\overset{\cdot}{C}OH + H^+$$

Naturally, it would be expected that loss of a proton would be slower than electron transfer and therefore rate-determining, e.g.

$$Mn(III) + CH_3OC_6H_4CH_3 \longrightarrow Mn(II) + CH_3OC_6H_4CH_3^+ \qquad \text{rapid}$$

$$CH_3OC_6H_4CH_3^+ \longrightarrow CH_3OC_6H_4CH_2^{\cdot} + H^+ \qquad \text{slow}$$

Variations on this basic mechanism arise in certain cases. Sometimes carbon—carbon bond fission is brought about in addition to proton loss, e.g.

$$Ce(IV) + CH_2(OH)CH(OH)CH_2(OH) \longrightarrow Ce(III) + CH_2O + H^+ + \overset{\cdot}{C}H(OH)CH_2(OH)$$

$$Mn(III) + (CH_3)_2\overset{\cdot}{C}HCOOH \longrightarrow Mn(II) + (CH_3)_2\overset{\cdot}{C}H + CO_2 + H^+$$

Frequently, complex formation between oxidant and substrate

takes place prior to the redox process, the latter then being referred to as inner-sphere in type. On occasions, these complexes are sufficiently long-lived to be discernible by colour and/or spectral changes as occur, for example, in the oxidation of methanol and ethanol by cerium(IV). Michaelis—Menten kinetics (first applied to enzyme reactions) are relevant in these cases. In the last few years, fast reaction techniques have been applied to reacting redox systems and there is an increasing amount of evidence for the existence of very short-lived intermediate complexes. An example is provided by

$$Mn^{3+} + (CH_3)_2COHCOOH \rightleftharpoons Mn^{3+} \cdot (CH_3)_2COHCOOH$$

$$Mn^{3+} \cdot (CH_3)_2COHCOOH \longrightarrow Mn(II) + (CH_3)_2\overset{\cdot}{C}OH + CO_2 + H^+$$

In the case of alcohols and carboxylic acids, equilibria of the type

$$ROH + Co^{III}(H_2O)_6^{3+} \rightleftharpoons R-O-Co^{III}(H_2O)_5^{2+} + H_3O^+$$

and

$$RCOOH + Co^{III}(H_2O)_6^{3+} \rightleftharpoons R-\underset{\underset{O}{\|}}{C}-O-Co^{III}(H_2O)_5^{2+} + H_3O^+$$

are thought to be involved in bringing the oxidisable substrate into close proximity with the oxidant thereby facilitating the transfer of a d electron with the production of Co(II) and either a RO· radical or a RCOO· radical. The size of the equilibrium constant dictates whether or not Michaelis—Menten kinetics are followed.

The substrate radical species formed as a result of the hydrogen atom abstractions outlined above may be removed from the system in a number of ways.

(a) By further oxidation, e.g.

$$\overset{\cdot}{C}HOHCH_2OH + Ce(IV) \longrightarrow 2\ CH_2O + H^+ + Ce(III)$$

$$C_6H_5CH_2^{\cdot} + Co(III) + H_2O \longrightarrow C_6H_5CH_2OH + H^+ + Co(II)$$

(b) by disproportionation, e.g.

$$C_6H_5\overset{\cdot}{C}HOCH_2C_6H_5 \longrightarrow C_6H_5CHO + C_6H_5CH_2^{\cdot}$$

(c) by coupling with another radical in the system, or

(d) by reaction with oxygen to give a peroxy radical. (This would correspond to the propagation of a chain mechanism and give rise to altered kinetics and products when the reaction is studied under anaerobic conditions.)

The presence of radicals may be inferred by their oxidation or reduction of added inorganic ions or by their ability to cause polymerisation to occur with added monomer, e.g. acrylamide or acrylonitrile. Furthermore, the nature of the radical may be assessed by end-group analysis of the resulting polymer species. Occasionally, the radical is present in sufficiently high concentrations to be detectable by electron spin resonance.

The principal concern of Chapter 3 is with reactions of the overall two-equivalent oxidants lead(IV) and thallium(III). Some reactions of mercury(II), palladium(II), and other metal ions are also considered. Of these oxidants, mercury(II), palladium(II), and thallium(III) are capable of accepting electrons from olefinic bonds to form π-complexes as a preliminary to oxidation. Lead(IV), in the form of lead tetraacetate, is probably the best-known oxidant after Cr(VI) and permanganate because of its usefulness in bringing about carbon—carbon bond fission in 1,2-diols and related compounds.

Considerable mechanistic complexity is met with in the reactions of the vigorous and much-used oxidants chromium(VI) and manganese(VII), which are the subject of Chapter 4. In the case of the former, an overall three-equivalent change takes place on reduction, the final state being Cr(III). Depending upon the reductant and the reaction conditions, various routes seem to be followed. Firstly, Cr(VI) can be reduced to Cr(III) in three consecutive one-equivalent stages. Of these, the conversion of Cr(V) to Cr(IV) is more likely to be rate-determining since it is probable that Cr(V) has a tetrahedral structure like Cr(VI) whereas Cr(IV) resembles the octahedral Cr(III) species. A second route is the two-equivalent reduction of Cr(VI) to Cr(IV) followed then by a one-equivalent reduction of Cr(IV) to Cr(III). Alternatively, an indirect conversion of Cr(IV) to Cr(III) may take place via Cr(V). Other possibilities exist. Permanganate may be reduced to the +2, +4, or +6 oxidation states, depending upon the nature of the reductant and on the pH of the reaction medium. As a general rule, an overall five-equivalent change may only be attained with certain active reductants in acidic media; more frequently, a net three-equivalent reduction occurs and MnO_2 is the product although strongly basic solution stabilises the manganate ion and reduction may be halted at this intermediate stage. The rate-determining step in the overall process is more often than not the direct reaction of Mn(VII) with the substrate to give Mn(VI). However, a two-equivalent reduction is sometimes encountered producing Mn(V) which in strongly basic solution can react with Mn(VII) to give Mn(VI) or, under less basic conditions, disproportionate to MnO_2 and Mn(VI).

Each chapter is prefixed by an account of the general features of the major oxidants discussed. Then follows a fairly extended summary of the reactions of the oxidants with other inorganic species. The oxidation of organic compounds is reviewed on the basis of substrate type. Details of reactions of some minor oxidants (not mentioned in the titles of the chapters) are included where appropriate.

Chapter 2

OXIDATIONS BY COBALT(III), CERIUM(IV), VANADIUM(V), MANGANESE(III), IRON(III), COPPER(II), AND SILVER(II)

1. GENERAL FEATURES OF THE OXIDANTS

(a) Cobalt(III)

Most kinetic studies have utilised dilute solutions of cobalt(III) prepared from cobalt(II) either by ozonolysis or by anodic oxidation. In a typical preparation [1], high-purity cobalt sponge (or, better, Co(II) perchlorate) dissolved in 6 M perchloric acid is oxidised at a low current density (~ 25 A m^{-2}) at 0°C using platinum electrodes. Maximum conversion ($\sim 70\%$) of Co(II) to Co(III) is achieved only after several hours. Concentrations of Co(III) perchlorate of up to 0.6 M have been attained by this method, the high solubility of the material preventing isolation of the solid salt [2]. Solutions are best prepared as required but they can be stored, if necessary, in a refrigerator for a few days. It should be noted that any marked deviation from the recommended procedures (for example, the use of more vigorous electrolysis or lower acidities) tends to produce solutions with different (and sometimes non-reproducible) kinetic and spectral properties. Acid solutions of cobalt(III) sulphate can be produced by an analogous method although, in view of the complexity of the equilibria involved in sulphate media, their use in kinetic studies gives rise to difficulties in interpretation. The use of solid cobalt(III) sulphate in sulphate or perchlorate media has the same drawback.

Cobalt(III) acetate in anhydrous acetic acid can be prepared electrolytically. Because of the low conductivity of this solvent, prolonged electrolysis is required (typically at 100 V, 50 mA for 1—2 days at room temperature) and even then only ~50% conversion takes place. The deep-green solution of Co(III) acetate can be purified by means of a cation-exchange resin (e.g. Amberlite IR-120H). The final solution (~10^{-3} M) is ~98% pure and is relatively stable.

Standard volumetric methods are readily applicable to the analysis of cobalt(III) solutions, e.g. back-titration of excess Fe(II) solution with Ce(IV) using ferroin indicator. Alternatively, the excess Fe(II) can be estimated spectrophotometrically. The absorption spectrum [1] of cobalt(III) in perchlorate solutions shows well-defined maxima at 400 nm and 605 nm (ϵ values of 4.1 and 3.5 m^2 mol^{-1}, respectively). With very dilute solutions, Co(III) can be monitored in the u.v. region (e.g. at 250 nm ϵ = 290 m^2 mol^{-1}). In acetic acid, Co(III) acetate has a shallow maximum at 630 nm.

The bulk of evidence indicates that a proportion of cobalt(III) in perchloric acid is dimeric and/or polymeric and that these polynuclear species are favoured by high concentrations of Co(III), low acidities, and high temperatures. Practical difficulties arising from the instability of perchlorate solutions of Co(III) have prevented the extent of polymerisation being established precisely but it appears likely that the dimeric species is formed from the monomeric Co^{3+} (i.e. $Co(H_2O)_6^{3+}$) ion via a hydrolysed species, viz.

$$Co^{3+} + H_2O \rightleftharpoons CoOH^{2+} + H^+ \quad \text{rapid, } K_h$$

$$Co^{3+} + CoOH^{2+} \longrightarrow Co-O-Co^{4+} + H^+ \quad \text{slow}$$

and/or

$$CoOH^{2+} + CoOH^{2+} \longrightarrow Co-O-CoOH^{3+} + H^+ \quad \text{slow}$$

followed, possibly, by the production of higher polymeric forms [3]. This scheme is in accord with the observation that polynuclear and monomeric forms are not readily interconvertible. For example, it seems that polymers persist when strong solutions of cobalt(III) are diluted. Potentiometric studies [4a, 4b] on solutions with total Co(III) concentrations of 10^{-3} M have indicated ~40% dimer for 0.5 M perchloric acid, falling to <10% for 3 M perchloric acid at the same temperature (23°C). At 1°C, cobalt(III) perchlorate solutions in 4 M perchloric acid are reported to be diamagnetic [5], an observation compatible with the low-spin (t_{2g}^6) electronic state for the monomeric $Co(H_2O)_6^{3+}$ ion. A kinetic examination [6] of the reduction of Co(III) by water has revealed the rate to be dependent on $[Co(III)]^{3/2}$ and on $[H^+]^{-2}$. This suggests that the rate-determining stage involves hydrolysed monomeric and dimeric species

$$CoOH^{2+} + HOCo-O-CoOH^{2+} \longrightarrow 3\ Co^{2+} + 2\ OH^- + HO_2^{\cdot}$$

Peroxy radicals are then oxidised further by

$$CoOH^{2+} + HO_2^{\cdot} \longrightarrow Co^{2+} + H_2O + O_2$$

There is considerable variation in the reported values for the hydrolysis constant of $Co(H_2O)_6^{3+}$, but it is likely to be ~2×10^{-3} mol l^{-1} at 25°C, increasing quite markedly with temperature ($\Delta H \sim +42$ kJ mol^{-1}) [7].

The electrode potential of the $Co^{3+}-Co^{2+}$ couple is reported to be +1.92 V at 25°C in 4 M perchloric acid and +1.85 V in the same concentration of nitric acid [4a, 8], the measured potentials being sensitive, as is usual, to changes in the complexing medium.

(b) Cerium(IV)

In perchloric acid media, cerium(IV) is best prepared by electrolytic oxidation of cerium(III) perchlorate [9]. Typically, 90% conversion is achieved and the metastable yellow cerium(IV) can be stored at 0°C in the absence of sunlight. The starting material is obtained from the metal or by evaporation of ammonium hexanitratocerate(IV) with concentrated hydrochloric acid followed by treatment with perchloric acid to remove chloride.

The electrode potential of the Ce(IV)—Ce(III) system is ligand-dependent, having a value of +1.70 V in 1 M perchloric acid and +1.44 V in 0.5 M sulphuric acid. Various equilibria complicate the solution chemistry of cerium(IV) in sulphate media [10].

$$Ce^{4+} + HSO_4^- \rightleftharpoons CeSO_4^{2+} + H^+ \qquad K_1 = 3500$$

$$CeSO_4^{2+} + HSO_4^- \rightleftharpoons Ce(SO_4)_2 + H^+ \qquad K_2 = 200$$

$$Ce(SO_4)_2 + HSO_4^- \rightleftharpoons Ce(SO_4)_3^{2-} + H^+ \qquad K_3 = 20$$

Formation of these complexes is a slow process; for example, when ammonium hexanitratocerate(IV) is dissolved in dilute sulphuric acid, the absorbance of the solution changes over several hours. For kinetic purposes, therefore, all sulphate solutions of cerium(IV) should be aged previous to use.

The electrode potential of the Ce(IV)—Ce(III) couple in sodium perchlorate/perchloric acid solutions varies with hydrogen ion concentration but is almost independent of perchlorate ion concentration [11]. This indicates that there is no substantial interaction between Ce(IV) or Ce(III) and the perchlorate ion and that hydrolysis of Ce(IV) takes place, viz.

$$Ce^{4+} + H_2O \rightleftharpoons CeOH^{3+} + H^+ \qquad K_h$$

$$CeOH^{3+} + H_2O \rightleftharpoons Ce(OH)_2^{2+} + H^+ \qquad K_h'$$

Dimerisation occurs by self-condensation of $CeOH^{3+}$ to give species of the type $CeOCe^{6+}$ and $HOCeOCeOH^{4+}$ [12a,12b]. Potentiometric studies [13] suggest that, in 2 M perchloric acid, the first hydrolysis step is at least 85% complete at 25°C ($K_h > 14$ mol l^{-1}). Since K_h' is small by comparison ($K_h' = 0.15$ mol l^{-1}), it is clear that $CeOH^{3+}$ is the predominant species present although the proportion of dimeric forms is difficult to evaluate. Evidently, further condensation to give polynuclear species takes place since solutions of cerium(IV) on prolonged standing show a colloidal deposit, particularly at low acidities.

Cerium(IV) solutions, if reasonably concentrated, can be titrated directly against standard Fe(II). More dilute solutions ($< 10^{-2}$ M) can be analysed by adding excess Fe(II) and then estimating the unreacted Fe(II) spectrophotometrically. For very dilute solutions, cerium(IV) can be monitored at 400 nm or at some convenient wavelength in the u.v. region.

(c) Vanadium(V)

In acidic solutions, vanadium can exist in four oxidation states, i.e. $V(H_2O)_6^{2+}$ (violet), $V(H_2O)_6^{3+}$ (dark blue), VO^{2+} (bright blue), and $V(V)$ (yellow). Since the oxidising power, as reflected by the redox potentials, increases regularly from the +2 state to the +5 state, each intermediate ion is stable to disproportionation.

Vanadium(V) in acid solution functions as a useful, if moderate, one-equivalent oxidant for organic compounds. The electrode potential for the system

$$VO_2^+ + 2 H^+ + e \rightleftharpoons VO^{2+} + H_2O$$

is $\sim +1$ V at $25°$C. In perchloric acid media, the dominant species is the oxocation which may be represented either as the simple (aquated) ion VO_2^+ or as the aquo ion $V(OH)_4^+$. In strongly acidic solution, there would appear to be kinetic evidence [14] for the existence of small concentrations of protonated forms of the type $V(OH)_3^{2+}$ resulting from the equilibria

$$VO_{2\,aq}^+ + H^+ \rightleftharpoons V(OH)_{3\,aq}^{2+}$$

and

$$V(OH)_{4\,aq}^+ + H^+ \rightleftharpoons V(OH)_{3\,aq}^{2+}$$

For sulphuric acid solutions, it has been suggested [15] that the following equilibrium is involved

$$(VO_2, 2H_2O)^+ + H_2SO_4 \rightleftharpoons (VO_2, H_2O, H_2SO_4)^+ + H_2O$$
$$(\text{or } V(OH)_4^+) \qquad\qquad\qquad (\text{or } V(OH)_3(HSO_4)^+)$$

Vanadium(V) can be prepared in a simple and direct way by dissolving sodium metavanadate in aqueous perchloric acid. Numerous oxidation studies have utilised sulphuric acid solutions, but here extensive complex formation takes place between vanadium(V) and sulphate ions.

Vanadium(V) can be analysed directly by spectrophotometry, or indirectly by reduction with sulphur dioxide followed by photometric determination (at 755 nm) of the vanadium (IV) produced. The exact stoichiometry of the reaction

$$V(V) + Fe(II) = V(IV) + Fe(III)$$

allows V(V) to be titrated directly with standard Fe(II) solution to a phenanthroline end-point.

(d) Manganese(III)

Until quite recently, there was little detailed information on the nature of the strongly oxidising manganese(III). The probable reason for this neglect lies in the tendency of Mn(III) to disproportionate

$$2\ Mn(III) \rightleftharpoons Mn(II)\ +\ Mn(IV)\,(as\ MnO_2)$$

However, there are satisfactory ways of avoiding this. The disproportionation can be reduced to a large extent by having a large excess of Mn(II) present. Manganese(III) can be stabilised further by employment of high acidities since disproportionation would seem to occur through the self-condensation of hydrolysed species. A third way, making use of strongly complexing media such as sulphate, oxalate or pyrophosphate, leads to obvious difficulties in the interpretation of kinetic data. The equilibrium constant for the disproportionation has been estimated to be $\sim 10^{-2}$ in 6 M perchloric acid [16] and $\sim 10^{-3}$ in 4 M sulphuric acid [17] at 25°C.

Manganese(III) solutions in aqueous perchloric acid (1—6 M) can be prepared [16], in concentrations up to $\sim 10^{-3}$ M, by electrolytic oxidation of a large excess of manganese(II) perchlorate at a platinum anode using a current density ~ 2 mA cm^{-2}. An alternative method is to add acidified permanganate solution to excess manganese(II) perchlorate in 4 M perchloric acid [18]. Formation of Mn(III) is slow, however, and increasing the temperature usually leads to the precipitation of manganese dioxide. Solid manganese(III) acetate can be made by the addition of potassium permanganate to a refluxing solution of anhydrous manganese(II) acetate in glacial acetic acid and allowing the mixture to cool while slowly adding water. The crude $Mn(OAc)_3.2H_2O$ is separated off and recrystallised from an acetic acid—water solvent [19].

Manganese(III) can be analysed by the addition of excess Fe(II), the unreacted Fe(II) being titrated with Ce(IV) using phenanthroline indicator. An alternative way is to estimate the Fe(III) formed spectrophotometrically at 260 nm. Direct monitoring of manganese(III) can be performed [16] in the visible region at 470 nm or in the u.v.

The effect of changes in acidity on the u.v. and visible spectrum of Mn(III) in perchloric acid solutions shows that hydrolysis takes place according to

$$Mn^{3+} + H_2O \rightleftharpoons MnOH^{2+} + H^+ \qquad K_h$$

Determination of the hydrolysis constant, K_h, is hampered to some extent by the instability of Mn(III) at low acidities even in the presence of a large excess of Mn(II). However, a careful study [20] has revealed a value for K_h of 0.93 mol l^{-1} at 25°C. The absorption spectra of the aquo and hydroxy complexes are given in Fig. 1; both show maxima at 470 nm although the extinction coefficient for Mn^{3+} is less than that of $MnOH^{2+}$. The cloudiness produced in Mn(III) solutions on prolonged standing has been ascribed [21] to the formation of dimeric species by condensation, leading to the production of higher (soluble) polymers and eventually to the precipitation of hydrated oxides, viz.

$$MnOH^{2+} + MnOH^{2+} \rightleftharpoons MnOMn^{4+} + H_2O$$

$$MnOMn^{4+} \rightleftharpoons Mn^{IV}O^{2+} + Mn^{2+}$$

$$dimer \longrightarrow soluble\ polymers \longrightarrow hydrated\ oxides$$

Electrode potentials of the Mn(III)—Mn(II) couple in various media are indicative of the degree of complexation. The values are +1.54 V for 3 M perchloric acid [22], +1.49 V for sulphate [23], and +1.15 V for pyrophosphate [24]. In this connection,

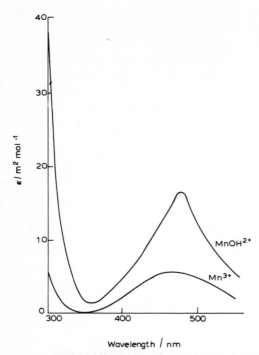

Wavelength / nm

Fig. 1. Absorption spectra of Mn^{3+} and $MnOH^{2+}$ at 25°C. (From Wells and Davies [20], by courtesy of The Chemical Society.)

it is interesting to note that, whereas hydrogen peroxide is rapidly attacked by Mn(III) in perchlorate media, peroxide is impervious to attack by Mn(III) strongly complexed with the substituted cyclohexanetetraacetate ligand [25].

(e) Iron(III) and copper(II)

For kinetic work, iron(III) solutions can be obtained by

oxidising, with a stoichiometric amount of hydrogen peroxide, acidic iron(II) perchlorate solution, itself prepared by addition of a slight excess of barium perchlorate solution to Fe(II) sulphate in perchloric acid followed by removal of the barium sulphate precipitate by filtration. Iron(III) solutions can be standardised by titration or, if very dilute, by spectrophotometry [20] (ϵ at 260 nm is 2.9×10^2 m^2 mol^{-1}). Common forms of iron(III) used in organic oxidations are the cyanide and phenanthroline complexes.

Copper(II) is a mild (and therefore specific) oxidant. Reduction leads usually to copper(I) although on occasions metallic copper may be formed as a result of disproportionation. There is no evidence to suggest that copper(III) is important as an intermediate unless strong oxidising agents, like peroxide, are present. An important application of copper(II) chloride and bromide in preparative organic chemistry is in oxidative halogenations of the type

$$2\ CuCl_2\ +\ CH_3COCH_3\ =\ 2\ CuCl\ +\ CH_3COCH_2Cl\ +\ HCl$$

(f) Silver(II)

Although the nature and reactivity of silver(II) was first appreciated in the 1930's, practical difficulties associated with handling solutions have tended to retard its development as an oxidant. Interest has revived, however, as a result of a detailed study reported [26] in 1963. Although little used at present, it would appear that silver(II) has considerable future scope as an oxidant for organic compounds since its electrode potential is very high ($\sim$ +2 V) and in terms of reactivity it is even more powerful than manganese(III). For this reason, an account is given of its main features.

Arising from an examination of the catalytic effect of silver (as silver(I)) on the redox reaction between cobalt(III) and

chromium(III), Sutcliffe and co-workers [27] have examined
the species of Ag(II) existing in perchlorate and nitrate media
and, in addition, have given an account of the factors governing
the instability of silver(II) in perchloric acid solutions.

Brown-coloured solutions of silver(II) can be prepared by
ozonolysis of silver(I), any trace impurities which might react
with silver(II) being oxidised in the process. Another method
is to use anodic oxidation of silver(I) solutions. (In a typical
preparation, 0.3 M $AgClO_4$ in 3 M $HClO_4$ is electrolysed at a
current of 40—100 mA using a rotating platinum anode and a
shielded cathode.) A third possibility is to employ silver(II)
oxide, prepared by the oxidation of silver(I) nitrate by
potassium peroxodisulphate ($K_2S_2O_8$) in alkaline solution, as
the starting material [28]. The black oxide is dissolved in
perchloric acid (typically 4 M) at -15 to $-20°C$ to produce a
dilute solution of Ag(II), the major amount of the solid being
reduced to Ag(I) on dissolution.

Silver(II) in solution has a distinctive absorption spectrum
[26] with a broad maximum at 475 nm ($\epsilon = 14$ m^2 mol^{-1},
resulting from an internal ligand field $d \leftarrow d$ transition), and
extending well into the visible region (shoulder at 575 nm). A
charge-transfer band, beginning at ~ 380 nm, increases rapidly
in intensity towards the u.v. region. The general insensitivity
of the spectrum to changes in perchloric acid concentration
suggests that there is little, if any, complexing between silver(II)
and perchlorate. In contrast, however, the spectrum of silver(II)
in nitric acid solutions is markedly dependent on acid concentration. Furthermore, there is a hypsochromic shift and
intensification of the absorption peak (to 380—400 nm,
$\epsilon_{max} \sim 220$ m^2 mol^{-1} for 6 M HNO_3) whilst the onset of the
charge-transfer band is also shifted to shorter wavelengths (to
330 nm). All these effects are indicative of extensive complex
formation between silver(II) and nitrate ions.

From measurements on the rate of disappearance of the

silver(II) absorption at 475 nm, the mechanism for the decomposition of silver(II) in perchlorate media is likely to be

$$2\ Ag^{2+} \rightleftharpoons Ag^+ + Ag^{3+} \qquad \text{rapid, } K_1$$

$$Ag^{3+} + H_2O \rightleftharpoons AgO^+ + 2\ H^+ \qquad \text{rapid, } K_2$$

$$AgO^+ \xrightarrow{k} Ag^+ + \tfrac{1}{2}\ O_2 \qquad \text{slow}$$

Assuming a steady-state concentration of the intermediate Ag(III) species, the derived rate law is

$$-\frac{d\left[Ag^{2+}\right]}{dt} = \frac{kK_1 K_2 \left[Ag^{2+}\right]^2}{\left[Ag^+\right]\left[H^+\right]^2}$$

in keeping with the observed kinetics. Although no quantitative information is available on the value of K_1, it is clear that the concentration of Ag(III) in equilibrium with Ag(I) and Ag(II) is very small.

The principal application of silver(II) in preparative organic chemistry at present is in the oxidation of primary and secondary amines.

2. REACTIONS WITH INORGANIC SPECIES

(a) Metal ions

Kinetic studies of oxidations of aquo metal ions by cobalt(III) are plentiful. The one-equivalent oxidations of iron(II) [29, 30], manganese(II) [16, 29], and chromium(II) [31] by cobalt(III) are rapid in perchloric acid. All three reactions follow the general rate law

$$-\frac{d\left[Co(III)\right]}{dt} = k'\left[Co(III)\right]\left[reductant\right]$$

with $k' = a + (b/[H^+])$ where a and b are constants. The obvious mechanism is

$$Co^{3+} \rightleftharpoons CoOH^{2+} + H^+ \qquad\qquad K_h$$

$$Co^{3+} + M^{2+} \xrightarrow{k_1} Co(II) + M(III)$$

$$CoOH^{2+} + M^{2+} \xrightarrow{k_2} Co(II) + M(III)$$

Thus a in the empirical rate law is equal to k_1 and b is k_2K_h. The corresponding oxidation of vanadium(II) by cobalt(III) is too rapid to study directly, even using stopped-flow techniques. The Co(III) + Cr(II) reaction is subject to catalysis by chloride ion and the products $CrCl^{2+}$, Cr^{3+}, and Co^{2+} have been identified by cation exchange (the product solution, on elution, gives three distinct bands coloured green, blue, and pink, respectively, with no indication of the dark-green chromium(III) dimer [31]). A flow technique [32a] has been used to demonstrate that the reaction between Co(III) and Fe(II) in the presence of chloride ion is of the inner-sphere type

$$CoCl^{2+} + Fe^{2+} \rightleftharpoons (CoClFe^{4+})^{\ddagger} \longrightarrow Co^{2+} + FeCl^{2+}$$

Mercury(I) is oxidised slowly by cobalt(III). Discounting the reaction between Co(III) and water, the stoichiometry is

$$2\ Co(III) + Hg(I)_2 = 2\ Co(II) + 2\ Hg(II)$$

Either of the following schemes is in accord with the kinetics [32b]

$$Co(III) + Hg(I)_2 \longrightarrow Co(II) + Hg(II) + Hg(I) \qquad \text{slow}$$

$$Co(III) + Hg(I) \longrightarrow Co(II) + Hg(II) \qquad\qquad \text{rapid}$$

or

$$Co(III) + Hg(I)_2 \longrightarrow Co(II) + Hg_2^{3+} \qquad \text{slow}$$

$$Co(III) + Hg_2^{3+} \longrightarrow Co(II) + 2\,Hg(II) \qquad \text{rapid}$$

In this system, the products of reaction, Co(II) and Hg(II), have no influence on the rate. However, in the oxidation of thallium(I) by cobalt(III), one of the products, Co(II), retards the reaction although the other, Tl(III), has no effect and the following scheme [33] is appropriate

$$Co(III) + Tl(I) \rightleftharpoons Co(II) + Tl(II)$$

$$Co(III) + Tl(II) \longrightarrow Co(II) + Tl(III)$$

The reaction between cobalt(III) and cerium(III) has received a detailed investigation by Sutcliffe and Weber [34a--34c]; particularly noteworthy is the attention given to the effect of anions on the rate. In perchlorate media, without other anions present, the rate-controlling step is the reaction between the hydroxo species of Co(III) and a perchlorate complex of Ce(III)

$$CoOH^{2+} + CeClO_4^{2+} \longrightarrow Co(II) + Ce(IV)$$

With nitrate and fluoride ions present, the slow stage becomes one between $CoOH^{2+}$ and $CeNO_3^{3+}$ or CeF^{2+}. In bisulphate media, sulphate complexes of both reactants participate in the reaction. Other metal ions oxidised by cobalt(III) include neptunium(V) [35] and vanadium(III) [32b].

Cerium(IV) in acidic sulphate media oxidises chromium(III) to chromium(VI)

$$3\,Ce(IV) + Cr(III) = 3\,Ce(III) + Cr(VI)$$

by a multi-stage process consisting of a succession of one-equivalent steps

$$Ce(IV) + Cr(III) \rightleftharpoons Ce(III) + Cr(IV)$$

$$Ce(IV) + Cr(IV) \rightleftharpoons Ce(III) + Cr(V) \quad \text{slow}$$

$$Ce(IV) + Cr(V) \rightleftharpoons Ce(III) + Cr(VI)$$

the second of which is rate-determining [36]. Another reaction studied in sulphuric acid media is that between manganese(II) and cerium(IV), the latter reacting in the form $Ce(SO_4)_2$ [37]. In perchloric acid, the kinetics of the reaction between iron(II) and cerium(IV) [9] disclose three possible routes involving the species Ce^{4+}, $CeOH^{3+}$, and $Ce(OH)_2^{2+}$. This oxidation is catalysed by HSO_4^- and F^-, but not by Cl^- ions; in the presence of added HSO_4^-, the reactive species is likely to be $CeSO_4^{2+}$ formed by

$$Ce^{4+} + HSO_4^- \rightleftharpoons CeSO_4^{2+} + H^+ \quad K = 3500$$

The rates of oxidation of Fe(II) (and tris(1,10-phenanthroline) Fe(II) complexes) by cerium(IV) have been employed by Dulz and Sutin [38] as a means of testing the linear free energy relationship predicted by Marcus [39]. The catalytic effect of iodide ions on the oxidation of arsenic(III) by cerium(IV), a well-known reaction used for the analysis of trace quantities of iodide, has been investigated [40] in sulphuric acid solution as also has catalysis of the same system by osmium [41]. The comparable cerium(IV)—antimony(III) system has been studied in perchloric acid media [42].

A few kinetic studies have been made of the reduction of vanadium(V) by other metal ions, for example

$$V(V) + Fe(II) = V(IV) + Fe(III) \qquad (\text{ref. } 43a, 43b)$$

$$V(V) + Ti(III) = V(IV) + Ti(IV) \qquad (\text{ref. } 44)$$

$$2 V(V) + Sn(II) = 2 V(IV) + Sn(IV) \qquad (\text{ref. } 45a, 45b)$$

In the V(V) + Fe(II) system, the complex acid dependence of the reaction rate suggests that at least three activation processes compete in the rate-determining stage, viz.

$$VO_2^+ + Fe^{2+} \longrightarrow (VO_2Fe^{3+})^{\ddagger}$$

$$VO_2^+ + Fe^{2+} + H^+ \longrightarrow (HVO_2Fe^{4+})^{\ddagger}$$

$$VO_2^+ + Fe^{2+} + 2 H^+ \longrightarrow (H_2VO_2Fe^{5+})^{\ddagger}$$

Oxidation of Fe(II), in the form $Fe(CN)_6^{4-}$, has also been reported [46] as has the interesting (if somewhat obscure) oxidation of the tantalum halide cluster ion [47], $Ta_6Br_{12}^{2+}$. In addition, details are available on the reactions of vanadium(V) with other vanadium ions, i.e. V(II) [48], V(III) [49], and V(IV) [50].

As noted earlier, interest in the reactions of manganese(III) is of recent origin. In the oxidations of vanadium(IV) [51] and iron(II) [16], a similarity in the hydrogen ion dependence of the reaction rates points to a common type of mechanism, e.g.

$$Mn^{3+} + VO^{2+} + H_2O \longrightarrow Mn^{2+} + VO_2^+ + 2 H^+$$

$$Mn^{3+} + H_2O \rightleftharpoons MnOH^{2+} + H^+$$

$$MnOH^{2+} + VO^{2+} \longrightarrow Mn^{2+} + VO_2^+ + H^+$$

In the oxidation of mercury(I) by manganese(III), the rate is reduced by both Hg(II) and Mn(II) [18]. The implication is that two rapidly established disproportionation equilibria precede two slow and rate-determining steps, viz.

$$2 \ Mn(III) \ \rightleftharpoons \ Mn(IV) \ + \ Mn(II) \qquad \text{rapid}$$

$$Hg(I)_2 \ \rightleftharpoons \ Hg(II) \ + \ Hg(0) \qquad \text{rapid}$$

$$Mn(III) + Hg(0) \longrightarrow Mn(II) \ + \ Hg(I) \qquad \text{slow}$$

$$Mn(IV) + Hg(I)_2 \longrightarrow Mn(II) \ + \ 2 \ Hg(II) \qquad \text{slow}$$

The step

$$Mn(III) + Hg(I) \longrightarrow Mn(II) + Hg(II) \qquad \text{rapid}$$

completes the sequence. Although the electrode potential of the Mn(III)—Mn(II) couple is considerably higher than that for Tl(III)—Tl(I) (+1.5 V and +1.2 V, respectively), manganese(III) will not oxidise thallium(I) in perchloric acid solution, even at a temperature of 80°C, unless chloride ions are present [52]. Rather surprisingly, although a precedent exists in the chloride-catalysed Ce(IV) + Tl(I) system, it appears that the reactive species is $\cdot Cl_2^-$ formed by

$$Mn^{3+} \ + \ Cl^- \ \rightleftharpoons \ MnCl^{2+}$$

$$MnCl^{2+} \ + \ Cl^- \ \rightleftharpoons \ Mn^{2+} \ + \ \cdot Cl_2^-$$

This, in turn, generates a Tl(II) intermediate

$$Tl(I) \ + \ \cdot Cl_2^- \longrightarrow Tl(II) \ + \ 2 \ Cl^-$$

which reacts further by

$$Tl(II) \ + \ \cdot Cl_2^- \longrightarrow Tl(III) \ + \ 2 \ Cl^-$$

Mechanistically, the oxidation of vanadium(III) by iron(III)

[53] is interesting in that the vanadium(IV) produced in the direct reaction

$$Fe(III) + V(III) \longrightarrow Fe(II) + V(IV)$$

is oxidised further by iron(III) to vanadium(V)

$$Fe(III) + V(IV) \rightleftharpoons Fe(II) + V(V)$$

which under the prevailing conditions is reduced again by reaction with Fe(II) or with V(III)

$$V(V) + V(III) \longrightarrow 2\ V(IV)$$

The Fe(III) + V(III) system is catalysed by copper(II) as shown by

$$Cu(II) + V(III) \longrightarrow Cu(I) + V(IV)$$
$$Fe(III) + Cu(I) \longrightarrow Fe(II) + Cu(II)$$

Oxidation of chromium(II) by iron(III) in perchloric acid solution is catalysed by chloride ion. From a detailed kinetic analysis, Dulz and Sutin [54] maintain, in principle, that there are two basic routes available to the oxidation, $Cr(H_2O)_5Cl^{2+}$ and $Fe(H_2O)_6^{2+}$ being the products in both, i.e. an *inner-sphere process*

$$Fe(H_2O)_5Cl^{2+} + Cr(H_2O)_6^{2+} \longrightarrow \left[(H_2O)_5Fe-Cl-Cr(H_2O)_5^{4+}\right]^{\ddagger}$$

and an *outer-sphere process*, represented either by

$$Fe(H_2O)_6^{3+} + Cr(H_2O)_5Cl^+ \longrightarrow \left[(H_2O)_5FeH_2O-Cl-Cr(H_2O)_5^{4+}\right]^{\ddagger}$$

or by

$$(H_2O)_6 Fe^{3+}Cl^- + Cr(H_2O)_6^{2+} \longrightarrow \left[(H_2O)_5 FeH_2O-Cl-Cr(H_2O)_5^{4+} \right]^{\ddagger}$$

In fact, the inner-sphere route is favoured in this case. The comparable reaction between $Fe(H_2O)_5Br^{2+}$ and Cr^{2+} has been examined [55a, 55b]. Similarly, iron(III), as $Fe(CN)_6^{3-}$, oxidises the pentacyano complex of cobalt(II), $Co(CN)_5^{3-}$, to yield the inert binuclear species $(NC)_5 Fe^{II}CNCo^{III}(CN)_5^{6-}$, indicating that a bridge (inner-sphere) mechanism is operating [56]. This is also the case in the reaction of $Fe(CN)_6^{3-}$ with $Cr(H_2O)_6^{2+}$ [57]. Other metal ions whose oxidations by iron(III) have been examined kinetically include vanadium(II) [58], tin(II) [59], europium(II) [60], uranium(IV) [61], neptunium(III) [62], neptunium(IV) [63], indium(I) [64], and copper(I) [65]. In the latter case the following inner-sphere scheme is plausible

$$Fe(H_2O)_6^{3+} \rightleftharpoons (H_2O)_5 FeOH^{2+} + H^+$$
$$(H_2O)_5 FeOH^{2+} + Cu^+ \longrightarrow Fe_{aq}^{2+} + CuOH^+ (\rightleftharpoons Cu_{aq}^{2+})$$

Copper(II) oxidises chromium(II) slowly in perchlorate media [66], the stoichiometric equation being simply

$$Cu(II) + Cr(II) = Cu(I) + Cr(III)$$

when Cu(II) is present in excess. However, when Cr(II) is in initial excess over Cu(II), metallic copper is rapidly produced by

$$Cu(I) + Cr(II) \longrightarrow Cu(0) + Cr(III)$$

and the net reaction then corresponds to

$$Cu(II) + 2 Cr(II) = Cu(0) + 2 Cr(III)$$

Similarly, copper(II) oxidises vanadium(II) in perchloric acid
solutions [67]. The rapid rate of the reaction is unexpected,
bearing in mind the generally held view that Cr(II) is much
more powerful than V(II) as a reductant. In hydrochloric
acid solution, it seems that copper(II) oxidises tin(II) via a
species of tin(III) [68]

$$Cu(II) \ + \ Sn(II) \ \rightleftharpoons \ Cu(I) \ + \ Sn(III)$$

$$Cu(II) \ + \ Sn(III) \ \longrightarrow \ Cu(I) \ + \ Sn(IV)$$

Copper(III), produced by pulse radiolysis of neutral aqueous
solutions of Cu(II) saturated with N_2O, is in equilibrium with
OH· radicals by

$$Cu^{2+} \ + \ OH\cdot \ \rightleftharpoons \ Cu(III) \ + \ OH^-$$

and the suggestion has been made [69] that OH· radicals are
the active species in those catalytic processes where Cu(III)
is suspected to be an intermediate [70].

Cobalt(III) oxidises chromium(III) very slowly in perchlorate
media but the rate is enhanced by the presence of silver(I) [27]

$$3 \ Co(III) \ + \ Cr(III) \ \overset{Ag(I)}{=\!=\!=} \ 3 \ Co(II) \ + \ Cr(VI)$$

The catalytic power of silver arises from the generation of a
silver(II) intermediate

$$Co(III) \ + \ Ag(I) \ \rightleftharpoons \ Co(II) \ + \ Ag(II)$$

which oxidises chromium(III) by a sequence of one-equivalent
stages, involving Cr(IV) and Cr(V), to the Cr(VI) final product.

$$Cr(III) \ + \ Ag(II) \ \longrightarrow \ Cr(IV) \ + \ Ag(I)$$

$$Cr(IV) \ + \ Ag(II) \ \longrightarrow \ Cr(V) \ + \ Ag(I)$$

$$Cr(V) \ + \ Ag(II) \ \longrightarrow \ Cr(VI) \ + \ Ag(I)$$

Silver(II) formed from Co(III) and Ag(I) in the absence of Cr(III) undergoes thermal decomposition to Ag(i) and oxygen (see p. 18).Transient amounts of Ag(II) in this system have been detected by spectrophotometric means and by electron spin resonance [71]. Stopped-flow techniques have been applied to study the catalysis of the Co(III)—Fe(II) system by silver(I) [8]. Here the silver(II) intermediate is in competition with Co(III) for Fe(II), the reactions being represented as

$$Co(III) + Ag(I) \rightleftharpoons Co(II) + Ag(II) \qquad K$$

$$Ag(II) + Fe(II) \longrightarrow Ag(I) + Fe(III)$$

$$Co(III) + Fe(II) \longrightarrow Co(II) + Fe(III)$$

A value of 4×10^{-2} for K is arrived at from the electrode potentials of the Co(III)—Co(II) and Ag(II)—Ag(I) couples. The cerium(IV) oxidations of mercury(I) and thallium(I) are both catalysed by silver(I) [72], e.g.

$$Ce(IV) + Ag(I) \rightleftharpoons Ce(III) + Ag(II)$$

$$Ag(II) + Tl(I) \longrightarrow Ag(I) + Tl(II)$$

$$Ce(IV) + Tl(II) \longrightarrow Ce(III) + Tl(III)$$

Here silver(II) and thallium(II) function as intermediates, the stoichiometry conforming to

$$2\ Ce(IV) + Tl(I) = 2\ Ce(III) + Tl(III)$$

As an oxidant, silver(II) is as much as 10^2 to 10^4 times more reactive than cobalt(III) towards vanadium(IV), chromium(III), manganese(II), iron(II), cerium(III), and cobalt(II). In addition, silver(I) is better than cobalt(II) as a reductant towards the oxidised forms of these ions. Evidently, silver(I) has con-

siderable potential as a redox catalyst and in this respect it
could prove superior to the more widely used cobalt(II) [8].

(b) Non-metallic species

Oxidation of chloride ion by cerium(IV) [73] and cobalt(III)
[74] has been studied kinetically. So also has oxidation of
bromide by cerium(IV) (in sulphate media) [75], cobalt(III)
[76a,76b], manganese(III) [77], and vanadium(V) [78]; and
iodide by cerium(IV) (in sulphate media) [75], cobalt(III)
[76b], vanadium(V) [79], and iron(III) [80a,80b]. In all
these studies, evidence is cited for the participation of halogen
atoms and/or radical ions. For example, in the case of the
Mn(III) + Br$^-$ system, two routes for the reaction are indicated
[77]. In the first, the initial metal—halide complex reacts with
Br$^-$ to give $\cdot$Br$_2^-$ which is then oxidised further to Br$_2$; in the
second, the complex decomposes directly to give bromine
atoms

$$Mn(III).Br^- + Br^- \longrightarrow Mn(II) + \cdot Br_2^- \qquad \text{slow}$$
$$Mn(III) + \cdot Br_2^- \longrightarrow Mn(II) + Br_2$$

or

$$Mn(III).Br^- \longrightarrow Mn(II) + Br\cdot \qquad \text{slow}$$
$$Br\cdot + Br\cdot \longrightarrow Br_2$$

In acid solution, oxidation of iodide ion by vanadium(V),
present as V(OH)$_4^+$, has the stoichiometry

$$2\ V(OH)_4^+ + 2\ I^- + 4\ H^+ = 2\ VO^{2+} + I_2 + 6\ H_2O$$

and it has been suggested that reaction takes place by attack
of V(OH)$_3^{2+}$, formed in a rapid pre-equilibrium, on iodide [81]

$$V(OH)_4^+ + H^+ \rightleftharpoons V(OH)_3^{2+} + H_2O \qquad\qquad \text{rapid}$$

$$V(OH)_3^{2+} + I^- \rightleftharpoons V(OH)_3I^+ \longrightarrow VOH^{2+} + OH^- + HIO$$

followed by conversion of V(III) and HIO to the final products, V(IV) and iodine, by

$$V(III) + V(V) \longrightarrow 2 V(IV)$$

$$HIO + I^- + H^+ \longrightarrow I_2 + H_2O$$

The oxidation of iodide ion by hexacyanoferrate(III) (ferricyanide)

$$2 Fe(CN)_6^{3-} + 2 I^- = 2 Fe(CN)_6^{4-} + I_2$$

is, rather unusually, a cation-catalysed reaction. For example, with K^+ ions present, the mechanism has been formulated [82] as

$$K^+ + Fe(CN)_6^{3-} \rightleftharpoons KFe(CN)_6^{2-} \qquad\qquad \text{(ion-pair)}$$

$$KFe(CN)_6^{2-} + I^- \rightleftharpoons IKFe(CN)_6^{3-}$$

$$IKFe(CN)_6^{3-} + I^- \longrightarrow KFe(CN)_6^{3-} + I_2^- \qquad\qquad \text{slow}$$

$$Fe(CN)_6^{3-} + I_2^- \longrightarrow Fe(CN)_6^{4-} + I_2$$

Thus the (initial) rate of this reaction is proportional to the first power of K^+ and $Fe(CN)_6^{3-}$ concentrations and to the square of the I^- concentration. Other halogen species, whose reactions have been studied kinetically, are bromine (to bromate by Co(III) [83]) and chlorine dioxide (to chlorate by Co(III) [84]).

30

In the rapid oxidation of hydrazoic acid by cobalt(III) in perchlorate media, viz.

$$2\ Co(III)\ +\ 2\ HN_3\ =\ 2\ Co(II)\ +\ 3\ N_2\ +\ 2\ H^+$$

participation of free radicals is indicated by the observation that reacting solutions quickly bring about polymerisation of acrylonitrile [85]. The reaction rate is first-order in both reactants and inversely proportional to hydrogen ion concentration [86a, 86b]. It is possible that an intermediate complex is formed which then decomposes by loss of a proton to give azide radicals [85]

$$Co(III)\ +\ HN_3\ \rightleftharpoons\ Co^{III}HN_3 \qquad \text{rapid}$$

$$Co^{III}HN_3\ \rightleftharpoons\ Co^{III}N_3^-\ +\ H^+ \qquad \text{rapid}$$

$$Co^{III}N_3^-\ \longrightarrow\ Co(II)\ +\ \cdot N_3 \qquad \text{slow}$$

$$2\ \cdot N_3\ \longrightarrow\ 3\ N_2 \qquad \text{slow}$$

Alternatively, the acid dependence may be attributed [86a, 86b] to the oxidation of HN_3 by $CoOH^{2+}$. Oxidation of hydrazoic acid by manganese(III), studied by a stopped-flow technique [87a, 87b], proceeds via the formation of a pink-coloured monoazido Mn(III) complex whose absorption spectrum and formation constant have been measured. Oxidation of the same substrate by cerium(IV) involves a similar type of complex, oscilloscope traces showing a rapid increase in absorbance (at 350 nm where Ce(III) and HN_3 do not interfere), followed by a slower decay [88].

The oxidation of sulphite by $Fe(CN)_6^{3-}$ has been the subject of much speculation. Originally, it was supposed that $\cdot SO_3^-$ radicals were the active intermediates, i.e.

$$Fe(CN)_6^{3-} + SO_3^{2-} \longrightarrow Fe(CN)_6^{4-} + \cdot SO_3^- \qquad \text{slow}$$

$$Fe(CN)_6^{3-} + \cdot SO_3^- + H_2O \longrightarrow Fe(CN)_6^{4-} + SO_4^{2-} + 2\,H^+ \qquad \text{rapid}$$

$$\cdot SO_3^- + \cdot SO_3^- \longrightarrow S_2O_6^{2-} \qquad \text{rapid}$$

but it has now been shown [89] that the reaction proceeds via more complex intermediates, viz.

$$Fe(CN)_6^{3-} + SO_3^{2-} \rightleftharpoons Fe(CN)_5(CNSO_3)^{5-}$$

$$Fe(CN)_5(CNSO_3)^{5-} + Fe(CN)_6^{3-} \rightleftharpoons Fe(CN)_5(CNSO_3)^{4-} + Fe(CN)_6^{4-}$$

$$Fe(CN)_5(CNSO_3)^{4-} + H_2O \longrightarrow Fe(CN)_6^{4-} + SO_4^{2-} + 2\,H^+$$

Thiocyanate is oxidised by Fe(III), viz.

$$6\,Fe(III) + SCN^- + 4\,H_2O = 6\,Fe(II) + CN^- + SO_4^{2-} + 8\,H^+$$

A mechanism compatible with the complex rate law is [90]

$$Fe^{3+} + SCN^- \rightleftharpoons FeSCN^{2+}$$

$$FeSCN^{2+} + SCN^- \rightleftharpoons Fe(SCN)_2^+$$

$$FeSCN^{2+} + SCN^- \longrightarrow Fe^{2+} + \cdot(SCN)_2^- \qquad \text{slow}$$

$$Fe(SCN)_2^+ + SCN^- \longrightarrow Fe^{2+} + \cdot(SCN)_2^- + SCN^- \qquad \text{slow}$$

$$Fe^{2+} + \cdot(SCN)_2^- \longrightarrow Fe^{3+} + 2\,SCN^-$$

$$Fe(III) + \cdot(SCN)_2^- \longrightarrow Fe^{2+} + (SCN)_2$$

$$(SCN)_2 + H_2O \longrightarrow CN^-, SO_4^{2-}, \quad \text{etc. as products}$$

Application of the steady-state approximation to the concentration of the intermediate $\cdot(SCN)_2^-$ gives rise to a rate expression in agreement with that experimentally determined. Other oxidations of iron(III) for which kinetic data are available are those with thiosulphate [91] and bisulphite [92].

Copper(II) in aqueous ammonia oxidises thiosulphate initially to tetrathionate. The following mechanism has been advanced [93].

$$Cu(NH_3)_4^{2+} + H_2O \rightleftharpoons Cu(NH_3)_3(H_2O)^{2+} + NH_3 \qquad \text{rapid}$$

$$Cu(NH_3)_3(H_2O)^{2+} + S_2O_3^{2-} \longrightarrow Cu(NH_3)_3(S_2O_3) + H_2O \qquad \text{slow}$$

$$Cu(NH_3)_3(S_2O_3) \longrightarrow Cu(NH_3)_3^+ + \cdot S_2O_3^- \qquad \text{rapid}$$

$$2 \cdot S_2O_3^- \longrightarrow S_4O_6^{2-} \qquad \text{rapid}$$

Silver(II) oxidises the dithionate ion to bisulphate

$$2\,Ag(II) + S_2O_6^{2-} + 2\,H_2O = 2\,Ag(I) + 2\,HSO_4^- + 2\,H^+$$

The rate law shows first-order dependences on Ag(II), $S_2O_6^{2-}$, and H^+ but the rate is independent of Ag(I) concentration [94]. A reasonable scheme is

$$H^+ + S_2O_6^{2-} \rightleftharpoons HS_2O_6^- \qquad \text{rapid}$$

$$Ag(II) + HS_2O_6^- + H_2O \longrightarrow Ag(I) + \cdot HSO_3 + HSO_4^- + H^+ \qquad \text{slow}$$

$$Ag(II) + \cdot HSO_3 + H_2O \longrightarrow Ag(I) + HSO_4^- + 2\,H^+ \qquad \text{rapid}$$

This contrasts with the oxidation of dithionate by chromium(VI) since there the rate is independent of the concentration of the oxidant and identical to that for hydrolysis of dithionate to sulphate and sulphite.

Phosphorous acid is oxidised by silver(II) in perchloric acid solution according to

$$2\,Ag(II) + H_3PO_3 + H_2O = 2\,Ag(I) + H_3PO_4 + 2\,H^+$$

The two-term rate law

$$-\frac{d\left[Ag(\text{II})\right]}{dt} = 2\,k_1\left[H_3PO_3\right] + \frac{2\,k_2\left[Ag(\text{II})\right]^2\left[H_3PO_3\right]}{\left[Ag(\text{I})\right]}$$

indicates two parallel paths [95]. The first term, in which k_1 increases with acidity, is in keeping with the dissociation of H^+ from a P—H bond producing an intermediate with a lone pair of electrons on the phosphorus atom, viz.

$$H_3PO_3 + H^+ \underset{\longleftarrow}{\longrightarrow} H_4PO_3^+ \qquad\qquad \text{rapid}$$

$$H_4PO_3^+ \underset{\longleftarrow}{\longrightarrow} (HO)_3P\text{:} + H^+ \qquad \text{slow}$$

$$(HO)_3P\text{:} + Ag(\text{II}) \longrightarrow \text{products} \qquad \text{rapid}$$

The value of k_1 agrees with the rate of exchange of deuterium between the P—H bond in the acid and D_2O solvent. The second term suggests two possibilities, either

$$2\,Ag(\text{II}) \underset{\longleftarrow}{\longrightarrow} Ag(\text{I}) + Ag(\text{III}) \qquad \text{rapid}$$

$$Ag(\text{III}) + H_3PO_3 + H_2O \longrightarrow Ag(\text{I}) + H_3PO_4 + 2\,H^+ \qquad \text{slow}$$

or

$$Ag(\text{II}) + P(\text{III}) \underset{\longleftarrow}{\longrightarrow} Ag(\text{I}) + P(\text{IV}) \qquad \text{rapid}$$

$$Ag(\text{II}) + P(\text{IV}) \longrightarrow Ag(\text{I}) + P(\text{V}) \qquad \text{slow}$$

Of these, the latter seems less likely. Oxidation of hypophosphorous acid by vanadium(V) in perchlorate media [96] is preceded by the formation of complexes of the type $VO_2^+.H_2PO_2^-$.

Copper(II) exerts a catalytic effect on the oxidation of molecular hydrogen by chromium(VI), the rate law being [97a, 97b]

$$-\frac{d\left[H_2\right]}{dt} = \frac{k_1 k_3\left[H_2\right]\left[Cu^{2+}\right]^2}{k_2\left[H^+\right] + k_3\left[Cu^{2+}\right]}$$

where the rate constants refer to the reactions

$$Cu^{2+} + H_2 \quad \underset{k_2}{\overset{k_1}{\rightleftharpoons}} \quad CuH^+ + H^+$$

$$CuH^+ + Cu^{2+} \quad \xrightarrow{k_3} \quad 2\ Cu^+ + H^+$$

Copper(II) is regenerated through

$$3\ Cu^+ + Cr(VI) = 3\ Cu^{2+} + Cr(III) \qquad \text{rapid}$$

Reversible heterolytic splitting of hydrogen also occurs with silver(I) as the catalyst [98].

There is strong evidence to suggest that oxidation of hydrogen peroxide by Co(III) [99], Ce(IV) [100a,100b], Mn(III) [101a,101b], Fe(III) [102], and Ag(II) [103] proceeds through the formation of intermediate metal ion—peroxide complexes of the type $M^{n+}-HO_2^-$. The stoichiometry corresponds to

$$2\ M^{n+} + H_2O_2 = 2\ M^{(n-1)+} + O_2 + 2\ H^+$$

and the reaction proceeds in two basic steps

$$M^{n+} + H_2O_2 \rightleftharpoons M^{(n-1)+} + HO_2^{\cdot} + H^+$$

$$M^{n+} + HO_2^{\cdot} \longrightarrow M^{(n-1)+} + O_2 + H^+$$

As an example, for the Co(III) + peroxide system in perchlorate media, Wells and Husain [99] have proposed the following comprehensive scheme involving, besides the $Co^{III}HO_2^-$ complex, a further complex $Co^{III}H_2O_2$.

$$\text{Co(III)} \rightleftharpoons \text{Co}^{III}\,\text{OH}^- + \text{H}^+$$

$$\text{Co(III)} + \text{H}_2\text{O}_2 \rightleftharpoons \text{Co}^{III}\,\text{H}_2\text{O}_2$$

$$\text{Co}^{III}\,\text{OH}^- + \text{H}_2\text{O}_2 \rightleftharpoons \text{Co}^{III}\,\text{HO}_2^-$$

$$\text{Co}^{III}\,\text{H}_2\text{O}_2 \rightleftharpoons \text{Co}^{III}\,\text{HO}_2^- + \text{H}^+$$

$$\text{Co}^{III}\,\text{H}_2\text{O}_2 \longrightarrow \text{Co(II)} + \text{HO}_2\cdot + \text{H}^+$$

$$\text{Co}^{III}\,\text{HO}_2^- \longrightarrow \text{Co(II)} + \text{HO}_2\cdot$$

$$\text{Co(III)} + \text{HO}_2\cdot \longrightarrow \text{Co(II)} + \text{O}_2 + \text{H}^+$$

where all the cobalt species are aquated. However, most of the Co(III) is complexed as $\text{Co}^{III}\text{HO}_2^-$ and decomposition of this complex is the major path in the overall reaction. Stopped-flow oscilloscope traces confirm the formation and subsequent breakdown of the complex. In the oxidation of hydrogen peroxide by cerium(IV), oxygen-18 experiments have demonstrated that all the evolved oxygen originates in the peroxide [104] and electron spin resonance has detected $\text{HO}_2\cdot$ as the radical intermediate [105]. In perchlorate media, the rate of decay of the intermediate complex (from measurements on stopped-flow traces) is independent of Ce(IV), peroxide, Ce(III), and acidity [100a], consistent with the scheme

$$\text{Ce(IV)} + \text{H}_2\text{O}_2 \rightleftharpoons \text{Ce}^{IV}\,\text{HO}_2^- + \text{H}^+$$

$$\text{CeOH}^{3+} + \text{H}_2\text{O}_2 \rightleftharpoons \text{Ce}^{IV}\,\text{HO}_2^-$$

$$\text{Ce}^{IV}\,\text{HO}_2^- \longrightarrow \text{Ce(III)} + \text{HO}_2\cdot \qquad \text{slow}$$

$$\text{Ce(IV)} + \text{HO}_2\cdot \longrightarrow \text{Ce(III)} + \text{H}^+ + \text{O}_2$$

although it has been pointed out that Wells and Husain may have followed the reaction between polymeric Ce(IV) and peroxide rather than that of the monomeric ions [106]. In sulphuric acid, the reaction is found to be retarded by the Ce(III) product. Evidently, the existence of a back-reaction involving Ce(III) is connected with the reduced redox potential of the Ce(IV)—Ce(III) couple in a sulphate as compared with a perchlorate medium (+1.4 V and +1.7 V, respectively). The manganese(III)—manganese(II) couple exerts a strong catalytic effect on the oxidation of hydrogen peroxide by chlorine in acidic media. This has been attributed [107] to the generation of free radicals in a reaction sequence of the type

$$Mn(III) + H_2O_2 \rightleftharpoons MnO_2H^{2+} + H^+$$

$$MnO_2H^{2+} + Cl_2 \longrightarrow Mn(II) + H^+ + Cl^- + O_2 + Cl\cdot$$

$$Mn(II) + Cl\cdot \longrightarrow Mn(III) + Cl^-$$

Manganese(III) oxidises hydroxylamine and nitrous acid according to [108]

$$6\ Mn(III) + NH_3OH^+ + 2\ H_2O = 6\ Mn(II) + NO_3^- + 8\ H^+$$

$$2\ Mn(III) + HNO_2 + H_2O = 2\ Mn(II) + NO_3^- + 3\ H^+$$

In perchlorate media, both are simple second-order reactions, showing an inverse dependence on hydrogen ion concentration. For the hydroxylamine reaction the first two steps in the suggested mechanism are

$$Mn^{3+} + NH_3OH^+ \longrightarrow Mn^{2+} + NH_2O\cdot + 2\ H^+$$

and

$$MnOH^{2+} + NH_3OH^+ \longrightarrow Mn^{2+} + NH_2O\cdot + H_3O^+$$

The intermediate radical is removed by oxidation, as represented by the net equation

$$NH_2O\cdot + 5\ Mn(III) + 2\ H_2O = NO_3^- + 5\ Mn(II) + 6\ H^+ \qquad rapid$$

rather than by dimerisation to give nitrogen

$$2\ NH_2O\cdot \longrightarrow N_2 + 2\ H_2O$$

In the faster reaction with nitrous acid, the sequence of steps is more complex

$$Mn^{3+} + HNO_2 \longrightarrow Mn^{2+} + H^+ + \cdot NO_2$$

$$MnOH^{2+} + HNO_2 \longrightarrow Mn^{2+} + H_2O + \cdot NO_2$$

$$Mn^{3+} + NO_2^- \longrightarrow Mn^{2+} + \cdot NO_2$$

$$MnOH^{2+} + NO_2^- \longrightarrow Mn^{2+} + \cdot NO_2 + OH^-$$

$$\cdot NO_2 + Mn(III) + H_2O = NO_3^- + Mn^{2+} + 2\ H^+ \qquad rapid$$

In neither reaction is there any retardation by added amounts of manganese(II) although this is a characteristic of the Mn(III) + HN$_3$ [87] and Mn(III) + H$_2$O$_2$ [101] systems. Evidently, the oxidising ability of the NH$_2$O· and ·NO$_2$ radicals is considerably less than that of the radicals encountered in the azide and peroxide reactions. Silver(II) oxidises hydroxylamine in a comparable way to manganese(III), the two rate-controlling steps being [109]

$$Ag^{2+} + NH_3OH^+ \xrightarrow{\ k_1\ } Ag^+ + NH_2O\cdot + 2\ H^+$$

and

$$AgOH^+ + NH_3OH^+ \xrightarrow{\ k_2\ } Ag^+ + NH_2O\cdot + H_3O^+$$

These are followed by the rapid oxidation of the NH$_2$O· radical to nitrate. The reaction is considerably faster than the oxidation by Mn(III). The observed first-order dependence on Ag(II), coupled with the lack of dependence on Ag(I), signifies that Ag(III) is not involved in the reaction. Hydroxylamine is oxidised by $Fe(CN)_6^{3-}$ in weakly acid solution [110] according to

$$Fe(CN)_6^{3-} + NH_2OH \longrightarrow Fe(CN)_6^{4-} + NH_2O\cdot + H^+$$

$$2 NH_2O\cdot \longrightarrow N_2 + 2 H_2O$$

A stopped-flow study has been made of the reaction of manganese(III) with hydrazine and methylhydrazines in perchloric acid solution, the rate of disappearance of the oxidant being followed at 470 or 510 nm [111]. In the case of hydrazine, the stoichiometry is

$$2 Mn(III) + 2 N_2H_5^+ = N_4H_8^{2+} \text{ (or } N_3H_5^{2+} + NH_4^+ \text{)} + 2 Mn(II) + 2 H^+ \text{ (or } + H^+\text{)}$$

and the rate of reaction is first-order in each reactant, varying inversely with acidity. The results suggest that oxidation proceeds through monoprotonated hydrazine species and protonated hydrazoyl radicals, $MnOH^{2+}$ reacting more quickly than Mn^{3+}. Methylation of hydrazine brings about a decrease in the rate of oxidation. In the reaction between cerium(IV) perchlorate and hydrazine the rate-determining stage has been identified as

$$CeOH^{3+} + N_2H_5^+ \longrightarrow Ce(III) + \cdot N_2H_4^+ + H_2O$$

although a different type of mechanism must apply since nitrogen gas is evolved with this oxidant but not with Mn(III) [112].

3. OXIDATION OF HYDROCARBONS

Hanotier et al. [113] have shown that the oxidising activity of cobalt(III) acetate in acetic acid is enhanced in the presence of strong acids with the result that aliphatic as well as aromatic hydrocarbons can be oxidised conveniently at low temperatures (20—40°C). They find that in pure acetic acid solvent, little if any reaction takes place with n-heptane (Bawn and Sharp found a similar lack of reactivity towards olefins, p. 45), but in the presence of trifluoroacetic acid (with oxygen absent) a fairly rapid reaction takes place, which is second-order in Co(III) concentration and is retarded by Co(II).

$$- \frac{d[Co(III)]}{dt} = \frac{k[RH][Co(III)]^2}{[Co(II)]}$$

However, it is noted that a concurrent direct reaction occurs with the solvent. The character of the products is dependent on whether oxygen is present or not; in the presence of oxygen, alcohols and ketones are formed, in its absence heptyl acetates result. Hanotier et al. conclude that the initial stage in the oxidation of the hydrocarbon by activated Co(III) is the formation of a radical, and that this process is reversible (thus explaining the retarding influence of Co(II)). The scheme

$$Co(III) + RH \underset{k_{-1}}{\overset{k_1}{\rightleftharpoons}} Co(II) + R\cdot$$

$$Co(III) + R\cdot \xrightarrow{k_2} products$$

and the corresponding rate law

$$- \frac{d[Co(III)]}{dt} = \frac{2k_1 k_2 [RH][Co(III)]^2}{k_2 [Co(III)] + k_{-1}[Co(II)]}$$

are compatible with the observed kinetics if

$$k_2 \left[Co(III)\right] \ll k_{-1} \left[Co(II)\right]$$

As regards products, the formation of heptyl acetates can be explained as arising from the reaction of carbonium ions with solvent.

$$Co(III) + R\cdot \longrightarrow Co(II) + R^+$$
$$R^+ + CH_3COOH \longrightarrow CH_3COOR + H^+$$

In the presence of oxygen, alkyl radicals might be expected to generate $RO_2^\cdot$ radicals. A chain reaction would then account for the formation of alcohols and ketones.

   Cobalt(III) oxidises benzene slowly in aqueous perchloric acid and the benzene ring is broken during the process. In contrast, manganese(III) is unreactive towards benzene and iron(III) also unless photo-activated, an indirect reaction then occurring through OH· radicals. The Co(III) oxidation conforms to a simple second-order rate law [114], the rate being directly proportional to the concentrations of both oxidant and benzene and showing the expected inverse dependence on acidity ($k = k_1 + (k_2/[H^+])$). Unlike the radical-induced decomposition of benzene, no biphenyl (or other hydrocarbon) is produced and oxygen has no effect on the products of reaction which are p-benzoquinone, muconic acid (HOOCCH=CHCH=CHCOOH) and its lactone. Mechanistically, it seems clear that the acid-independent and acid-dependent paths are, respectively

$$Co^{3+} + C_6H_6 \longrightarrow Co(II) + C_6H_5\cdot + H^+ \qquad \text{slow}$$

and

$$CoOH^{2+} + C_6H_6 \longrightarrow Co(II) + C_6H_5\cdot + H_2O \qquad \text{slow}$$

assuming, for simplicity, the monomeric form of Co(III). The phenyl radical is then consumed by

$$Co^{3+}_{aq} + C_6H_5{\cdot} \longrightarrow Co(II) + C_6H_5OH + H^+ \qquad \text{rapid}$$

and

$$CoOH^{2+} + C_6H_5{\cdot} \longrightarrow Co(II) + C_6H_5OH \qquad \text{rapid}$$

Competition from

$$C_6H_5{\cdot} + O_2 \longrightarrow C_6H_5O_2{\cdot}$$

does not take place. If, under the conditions of the kinetics, Co(III) is dimeric, then oxidation to phenol will occur in one step rather than by the two as written. The overall course of the reaction can be represented in terms of the sequences: (a) benzene → phenol → hydroquinone → p-benzoquinone, and (b) benzene → phenol → catechol → o-benzoquinone → muconic acid (+ lactone).

Cooper and Waters [115] have found the order of reactivity of aromatic hydrocarbons and of cyclohexane towards cobalt(III) perchlorate in aqueous methyl cyanide solutions to be anthracene ~ phenanthrene > naphthalene > biphenyl > p-$tert.$-butyltoluene > diphenylmethane > ethylbenzene ~ toluene ~ bibenzyl > cyclohexane > p-nitrotoluene. Toluene is oxidised to the side-chain products, benzyl alcohol, benzaldehyde, and benzoic acid, the product distribution (and therefore the overall stoichiometry) varying with the concentration of toluene and with the composition of the solvent. Cyclohexane is oxidised to cyclohexanol and cyclohexanone and then further to adipic acid. Cooper and Waters conclude that the aromatic hydrocarbons react by direct transfer of a π electron to Co(III) whereas the aralkyl hydrocarbons (e.g.

toluene and ethylbenzene) react by transferring a hydrogen atom from an aliphatic C—H group.

Alkylbenzenes are oxidised by cobalt(III) acetate in acetic acid, the rate being increased by the addition of acetate and decreased by the addition of cobalt(II) [116a,116b]. The reactivity order (in the presence of potassium acetate) is p-xylene > toluene ~ ethylbenzene > cumene ~ p-chlorotoluene. At the temperature employed (65°C) thermal self-decomposition of Co(III) acetate is significant. In the presence of acetate, the latter reaction yields carbon dioxide and methane as products and, like the hydrocarbon oxidations, is greatly accelerated by the addition of lithium chloride. These results, together with the observation that chloride ion has a marked effect on the spectrum of Co(III) acetate, indicate the formation of a reactive chloride complex. In the absence of chloride, the principal products resulting from the oxidation of toluene are benzyl acetate and benzaldehyde; in the presence of chloride, benzyl chloride and chlorotoluene (in lesser amounts) are also formed. A basic difference between these Co(III) oxidations and those by Mn(III) [117] and Pb(IV) [118] acetates is that in the latter two cases, a —CH₂OAc adduct (methylbenzyl acetate) occurs as a result of a radical mechanism involving the ·CH₂COOH species, formed by thermal breakdown of the oxidant (see pp. 74 and 137). For the Co(III) case, Heiba et al. [116a] suggest the electron transfer mechanism

Evidence for radical cation intermediates of this type has been obtained from spectrophotometry and electron spin resonance. Nucleophilic attack on the ring accounts for the formation of p-chlorotoluene in the presence of chloride ions.

In the absence of oxygen, the oxidation of ethylbenzene by cobalt(III) acetate ceases when the concentrations of Co(III) and Co(II) become equal [119]. This suggests that a mixed Co(III)—Co(II) dimer is formed which is unreactive, the reactive form being a Co(III)—Co(III) dimer. It has been proposed that an intermediate benzylic radical is produced in the initial step and is oxidised to the corresponding cation which then reacts with acetic acid solvent to yield an acetate product. In the presence of oxygen, however, the radical intermediate reacts by

$$C_6H_5\dot{C}HCH_3 + O_2 \longrightarrow C_6H_5CH(\dot{O}_2)CH_3$$

and acetophenone is the product.

In the presence of strong acids (e.g. H_2SO_4), alkylbenzenes and other alkylarenes are readily oxidised by manganese(III) acetate (at reaction temperatures as low as 20—40°C), the products arising solely from side-chain oxidation [120]. The kinetics bear a close resemblance to those for the reaction of p-methoxytoluene with Mn(III) [19] at 70—100°C. The suggested mechanism is similar to that for the Co(III) + ethylbenzene system and a mixed Mn(II)—Mn(III) complex may well be involved.

Toluene is very much more reactive towards vanadium(V) than its m- or p-nitro derivatives in acetic acid—sulphuric acid mixtures, the rate of oxidation increasing with acidity [121]. Cerium(IV) reacts at lower temperatures and acidities than does V(V) and the orders of reactivity towards this oxidant in aqueous acetic acid—sulphuric acid media are toluene > m-nitrotoluene ≫ p-nitrotoluene and p-chlorotoluene > o-chlorotoluene > m-chlorotoluene [122]. The major product in each case is the aldehyde.

In dilute sulphuric acid, cobalt(III) sulphate oxidises unsaturated hydrocarbons (to a complex mixture of products) in a simple second-order manner [123a, 123b]

$$-\frac{d[Co(III)]}{dt} = k[Co(III)][RH]$$

which suggests the initial step

$$Co(III) + RCH = CH_2 \longrightarrow Co(II) + R\dot{C}H - \overset{+}{C}H_2$$

Care was taken by Bawn and Sharp to purify the olefins (e.g. pentene, hexene, octene) from traces of peroxide impurities since these are highly reactive towards Co(III). In sulphate media Co(III) is so strongly complexed that reaction with water is completely suppressed. When glacial acetic acid containing sulphuric acid is used as a medium, the rate of disappearance of Co(III) becomes simply

$$-\frac{d\left[Co(III)\right]}{dt} = k\left[Co(III)\right]$$

and is independent of both olefin and hydrogen ion concentrations. However, if Co(III) sulphate in sulphuric acid is added to an acetic acid solution of the olefin, the reaction rate and kinetics are similar to those for aqueous solutions. Olefins in pure acetic acid are not attacked by Co(III).

4. OXIDATION OF ALCOHOLS

Apart from iron(III), all the oxidants discussed in this chapter oxidise primary and secondary alcohols to aldehydes and ketones, respectively. Formation of complexes between substrate and oxidant is common and Michaelis—Menten kinetics apply in a number of cases. Tertiary alcohols are attacked only by cobalt(III).

In aqueous sulphuric acid, cobalt(III) oxidises methyl, ethyl, and n-propyl alcohols in such a way as to lead Bawn and White to suggest that the initial step is the production, by electron-transfer, of a RO· radical [124]. The radical is then supposed

46

to react further with solvent to produce an OH· radical, or be oxidised irreversibly by Co(III). In the case of ethanol, the proposed steps are

$$Co(III) + C_2H_5OH \longrightarrow Co(II) + C_2H_5O\cdot + H^+$$
$$C_2H_5O\cdot + H_2O \rightleftharpoons C_2H_5OH + OH\cdot$$
$$Co(III) + C_2H_5O\cdot \longrightarrow Co(II) + CH_3CHO + H^+$$

These are followed by further steps in which acetaldehyde is oxidised to acetic acid, initiated by

$$Co(III) + CH_3CHO \longrightarrow Co(II) + CH_3CO\cdot + H^+$$

Ardon [125] has made a kinetic study of the oxidation of ethanol by cerium(IV) in perchloric acid (the oxidation of methanol has also been investigated [126]). It is well known that addition of an alcohol to a cerium(IV) solution gives rise to a marked deepening in colour from yellow to red, an effect forming the basis of a colorimetric method for the analysis of alcohols. Once formed, the Ce(IV)—alcohol complex is unstable. With alcohol in excess, the rate of disappearance of Ce(IV) follows a first-order rate law

$$-\frac{d[Ce(IV)]_t}{dt} = k_{obs}[Ce(IV)]_t$$

where $[Ce(IV)]_t$ represents the total concentration of Ce(IV) present (i.e. both complexed and uncomplexed). On the basis of the mechanism

$$Ce(IV) + ROH \rightleftharpoons Ce(IV)\cdot ROH \qquad K$$
$$Ce(IV)\cdot ROH \xrightarrow{k} products$$
$$-\frac{d[Ce(IV)]_t}{dt} = kK[Ce(IV)][ROH]$$

where $[Ce(IV)]$ is the concentration of uncomplexed $Ce(IV)$.
Since

$$\underset{\text{(uncomplexed)}}{[Ce(IV)]} + \underset{\text{(complexed)}}{K[Ce(IV)][ROH]} = \underset{\text{(total)}}{[Ce(IV)]_t}$$

then

$$[Ce(IV)] = \frac{[Ce(IV)]_t}{1 + K[ROH]}$$

and the derived rate law becomes

$$-\frac{d[Ce(IV)]_t}{dt} = \frac{kK[Ce(IV)]_t[ROH]}{1 + K[ROH]}$$

Thus the observed rate constant is identified as

$$k_{obs} = \frac{kK[ROH]}{1 + K[ROH]}$$

and it follows that

$$\frac{1}{k_{obs}} = \frac{1 + K[ROH]}{kK[ROH]} = \frac{1}{kK[ROH]} + \frac{1}{k}$$

The mechanism is validated by the observed linearity of plots
of $1/k_{obs}$ versus $1/[ROH]$ (see Fig. 2). Evaluation of the
slopes ($= 1/kK$) and intercepts ($= 1/k$) of such plots allows
values for k and K to be calculated. At 20°C for 3.2 M perchloric acid, Ardon found $k = 7 \times 10^{-3}$ s^{-1} and $K = 4.3$ l mol^{-1}.
The latter value has been verified by a spectrophotometric
study of the changes in absorbance brought about by varying
amounts of alcohol at low $Ce(IV)$ concentrations where the
rate of oxidation is low. Although it is assumed in the above

48

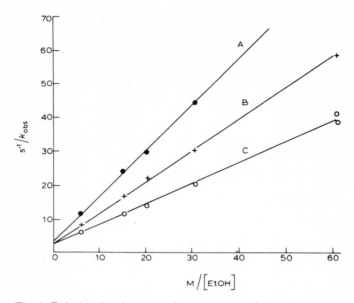

Fig. 2. Relationship between the reciprocal of the observed rate constant and the reciprocal of the alcohol concentration for the Ce(IV) + ethanol reaction at $HClO_4$ concentrations of A, 0.79 M; B, 1.73 M; and C, 3.20 M at 20°C and constant ionic strength. (From Ardon [125], by courtesy of The Chemical Society.)

that a 1:1 Ce(IV)—alcohol complex is involved, it should be noted that higher complexes may well be formed at greater alcohol concentrations.

In contrast to its outer-sphere oxidation of secondary alcohols, manganese(III) oxidises primary alcohols (as instanced by methanol)

$$2 \, Mn(III) \; + \; CH_3OH \; = \; 2 \, Mn(II) \; + \; HCHO \; + \; 2 \, H^+$$

through an intervening metal ion—alcohol complex and the reaction is inner-sphere in type [127]. The rate of reaction is first-order in both total Mn(III) and CH_3OH concentrations and is acid-dependent. The retarding influence of Mn(II) cannot be explained on the basis of the disproportionation

$$2\ Mn(III) \rightleftharpoons Mn(II) + Mn(IV)$$

since the rate would be second-order in total Mn(III). Wells and Barnes [128] conclude that the basis of the mechanism is the following sequence of reactions.

$$Mn^{3+} + H_2O \rightleftharpoons MnOH^{2+} + H^+$$
$$CH_3OH + H^+ \rightleftharpoons CH_3\overset{+}{O}H_2$$
$$Mn^{3+} + CH_3OH \rightleftharpoons Mn^{3+} \cdot CH_3OH$$
$$Mn^{3+} \cdot CH_3OH \rightleftharpoons Mn^{3+} \cdot CH_3O^- + H^+$$
$$Mn^{3+} \cdot CH_3OH \rightleftharpoons (Mn^{2+} \cdot H^+ CH_2OH) \longrightarrow Mn^{2+} + \cdot CH_2OH + H^+$$
$$Mn^{3+} \cdot CH_3O^- \rightleftharpoons (Mn^{2+} \cdot CH_2OH) \longrightarrow Mn^{2+} + \cdot CH_2OH$$
$$Mn(III) + \cdot CH_2OH \longrightarrow Mn(II) + HCHO + H^+$$

In this scheme, it is assumed that break-down of the $Mn^{3+} \cdot CH_3OH$ and $Mn^{3+} \cdot CH_3O^-$ complexes produces initially Mn^{2+}, H^+ and $\cdot CH_2OH$ radicals trapped in a solvent cage and that the effect of added Mn(II) is to bring about a back reaction by increasing the proportion of Mn^{2+} in the cage. Similar complexes are formed by vanadium(V) [14].

In some instances, it would appear that a radical is formed by fission of a C—C bond rather than a C—H bond [129]. For example, vanadium(V) oxidises β-phenyl ethanol to benzyl alcohol and then to benzaldehyde in the sequence

$$C_6H_5CH_2CH_2OH \longrightarrow C_6H_5CH_2 \cdot (+ CH_2O) \longrightarrow C_6H_5CH_2OH \longrightarrow C_6H_5CHOH \longrightarrow C_6H_5CHO$$

rather than to phenylacetic acid by

$$C_6H_5CH_2CH_2OH \longrightarrow C_6H_5CH_2CHOH \longrightarrow C_6H_5CH_2CHO \longrightarrow C_6H_5CH_2COOH$$

Also, *tert.*-butyl benzyl alcohol reacts with V(V) to give benzaldehyde as the main product, presumably by loss of a *tert.*-butyl radical

$$C_6H_5CH(OH)C(CH_3)_3 \longrightarrow C_6H_5CHO + \cdot C(CH_3)_3$$

The oxidation of isopropanol by Ce(IV) in perchlorate media conforms to the stoichiometry

$$2\, Ce(IV) + (CH_3)_2CHOH = 2\, Ce(III) + (CH_3)_2CO + 2\, H^+$$

Variation in the intercept and slope of $1/k_{obs}$ versus $1/[ROH]$ plots with acidity indicates that two intermediate complexes are formed, $Ce^{4+}\cdot(CH_3)_2CHOH$ and $Ce^{4+}\cdot(CH_3)_2CHO^-$, the two forming an acid—base equilibrium. The essentials of the scheme suggested by Wells and Husain [130] are

$$Ce^{4+} + H_2O \rightleftharpoons CeOH^{3+} + H^+$$
$$Ce^{4+} + (CH_3)_2CHOH \rightleftharpoons Ce^{4+}\cdot(CH_3)_2CHOH \qquad K_1$$
$$Ce^{4+}\cdot(CH_3)_2CHOH \rightleftharpoons Ce^{4+}\cdot(CH_3)_2CHO^- + H^+ \qquad K_2$$
$$(CH_3)_2CHOH + H^+ \rightleftharpoons (CH_3)_2CHO^+H_2$$
$$Ce^{4+}\cdot(CH_3)_2CHOH \xrightarrow{k_1} Ce(III) + (CH_3)_2\dot{C}OH + H^+ \qquad slow$$
$$Ce^{4+}\cdot(CH_3)_2CHO^- \xrightarrow{k_2} Ce(III) + (CH_3)_2\dot{C}OH \qquad slow$$
$$Ce(IV) + (CH_3)_2\dot{C}OH \longrightarrow Ce(III) + (CH_3)_2CO + H^+$$

At 25°C and a constant ionic strength of 2.06 M, values for the equilibrium and rate constants in the above scheme are $K_1 = 10\ l\ mol^{-1}$, $K_2 = 2.1\ mol\ l^{-1}$, $k_1 = 2.2 \times 10^{-3}\ s^{-1}$, and $k_2 = 1.02 \times 10^{-3}\ s^{-1}$. *Sec.*-butanol and cyclohexanol are oxidised by Ce(IV) via an inner-sphere process similar to that for isopropanol [131a,131b]. At 25°C the rate constants are $k_1 = 2.3 \times 10^{-3}\ s^{-1}$ and $k_2 = 1.49 \times 10^{-3}\ s^{-1}$ (*sec.*-butanol); $k_1 = 8.9 \times 10^{-3}\ s^{-1}$ and $k_2 = 3.0 \times 10^{-3}\ s^{-1}$ (cyclohexanol). Although both k_1 and k_2 increase in the order isopropanol <

sec.-butanol < cyclohexanol, it is interesting that the activation energies for the corresponding steps do *not* decrease in the order isopropanol > *sec.*-butanol > cyclohexanol.

Oxidation of isopropanol, *sec.*-butanol and cyclohexanol by manganese(III) takes place without formation of an intermediate Mn(III)—alcohol complex [132a,132b], for example

$$Mn^{3+} + H_2O \rightleftharpoons MnOH^{2+} + H^+$$
$$(CH_3)_2CHOH + H^+ \rightleftharpoons (CH_3)_2CH\overset{+}{O}H_2$$
$$Mn^{3+} + (CH_3)_2CHOH \longrightarrow Mn(II) + (CH_3)_2\dot{C}OH + H^+$$
$$MnOH^{2+} + (CH_3)_2CHOH \longrightarrow Mn(II) + (CH_3)_2\dot{C}OH + H_2O$$
$$Mn(III) + (CH_3)_2\dot{C}OH \longrightarrow Mn(II) + (CH_3)_2CO + H^+$$

The overall reaction has a stoichiometry given by

$$2\,Mn(III) + (CH_3)_2CHOH = 2\,Mn(II) + (CH_3)_2CO + 2\,H^+$$

The rate for all three alcohols is first-order in both alcohol and Mn(III) and is independent of hydrogen ion concentration. Radical participation is indicated by the polymerisation produced in acrylonitrile. The reaction between the hexaaquomanganese(III) ion and the unprotonated alcohol molecule is rate-controlling. Values of the rate constant for this step increase in the order isopropanol < *sec.*-butanol < cyclohexanol $(4.9 \times 10^{-5}, 8.9 \times 10^{-5}$ and 2.4×10^{-4} l mol^{-1} s^{-1} at 25°C) whereas activation energies increase in the reverse order. These observations can be rationalised in terms of inductive effects and electron availability on the $\diagup$CHOH group.

Cyclohexanol is oxidised to adipic acid by vanadium(V) in perchloric acid solution through an intervening $V(OH)_3^{2+}$. ROH complex which is reddish in colour [14]. During the course of the reaction, the red colour persists until it becomes obscured by the blue colour of the vanadium(IV) product. Substitution of 1-deuterocyclohexanol brings about a kinetic isotope effect which suggests that a C—H bond is broken in the rate-controlling stage. In sulphuric acid media, the scheme is modified to

$$V(OH)_3^{2+} + HSO_4^- \rightleftharpoons V(OH)_3HSO_4^+$$

$$V(OH)_3HSO_4^+ + ROH \overset{fast}{\rightleftharpoons} [complex]^+ \overset{slow}{\longrightarrow} products$$

In contrast to the oxidation of cyclohexanone which is rapid, complete oxidation of cyclohexanol requires up to five days at 50°C in 4.2 M sulphuric acid and it is possible that a resistant intermediate, such as cyclohexyl adipate, is formed.

It has been suggested that the oxidation of cyclobutanol may be used as a means of diagnosing one-equivalent and two-equivalent oxidants [133]. Reaction with Cr(VI) to give Cr(IV) yields the cyclic ketone whereas the one-equivalent oxidants, Mn(III), V(V), Ce(IV) (and Cr(IV)), produce the acyclic hydroxyaldehyde, $HO(CH_2)_3CHO$, by cleavage of the carbon—carbon bond, viz.

Unlike most other oxidants, cobalt(III) is capable of oxidising tertiary alcohols. In perchlorate solutions, the oxidation of *tert.*-butyl alcohol is accomplished, according to Hoare and Waters [134a, 134b], in a series of stages

$$Co(III) + (CH_3)_3COH \rightleftharpoons Co(III) \cdot (CH_3)_3COH$$

$$Co(III) \cdot (CH_3)_3COH \longrightarrow (CH_3)_3CO \cdot + Co(II) + H^+ \quad slow$$

$$(CH_3)_3CO \cdot \longrightarrow (CH_3)_2C = O + CH_3$$

$$Co(III) + CH_3 + H_2O \longrightarrow CH_3OH + Co(II) + H^+$$

Methanol is then oxidised further to formic acid as represented by

$$2 \text{ Co(III)} + CH_3OH \longrightarrow 2 \text{ Co(II)} + CH_2(OH)_2$$
$$2 \text{ Co(III)} + CH_2(OH)_2 \longrightarrow 2 \text{ Co(II)} + HCOOH$$

Kinetically, the reaction is simple second-order showing the inverse dependence on hydrogen ion concentration so typical of reactions of Co(III). Rather unexpectedly, the higher tertiary alcohols are oxidised more easily than *tert.*-butyl alcohol. Their oxidations have also been examined but, for reasons of solubility, a mixed solvent of methyl cyanide and water was employed.

The rate of oxidation of allyl and crotyl alcohols by V(V) is much greater than that of comparable primary or secondary saturated alcohols and Jones and Waters suggest that the radical formed in the rate-determining step is stabilised in the former systems [135]. Additional support for this suggestion comes from the observation that allyl alcohol reacts with Mn(III) pyrophosphate whereas saturated alcohols do not [136].

5. OXIDATION OF GLYCOLS

Glycols are oxidised readily by cerium(IV), vanadium(V), and manganese(III) and a considerable number of kinetic investigations have been reported. In the case of vicinal glycols, cyclic complexes play an important role in the oxidation sequence. Much attention has been directed to determining whether C—C or C—H bond fission occurs. Cobalt(III) is capable of oxidising certain glycols but there is a surprising lack of kinetic information.

In perchlorate media, glycerol is oxidised by cerium(IV)

through a 1:1 Ce(IV)—glycerol complex [137].

$$\text{Ce(IV)} + \text{glycerol} \underset{}{\overset{K}{\rightleftharpoons}} \text{complex} \xrightarrow{k} \text{products}$$

The complex has been identified both spectrophotometrically and kinetically (K = 19.8 l mol^{-1} and 20.0 l mol^{-1}, respectively, at 20°C). In sulphuric acid, the rate-determining step could involve the breaking of a C—C bond, viz.

Addition of acrylamide brings about a reduction in the rate of oxidation, an indication of the presence of intermediate radicals. For perchloric acid solutions, there is evidence to suggest that more than one intermediate glycol—Ce(IV) complex is formed in the butane-2,3-diol system [138]. In sulphuric acid solution, cerium(IV) is already strongly complexed with sulphate ions and has little, if any, ability to coordinate with the substrate and no complex formation is detected for both butane-2,3-diol and glycerol. This is not the case, however, in the cerium(IV) sulphate oxidation of ethandiol [139] and alcohols; evidently, the type of hydroxy compound is important, too, in deciding the tendency to complex.

Supporting evidence for the existence of intermediate radicals in the oxidation of glycols by cerium(IV) is the report that the yield of cyclohexanone, resulting from the oxidation of bicyclohexyl-1, 1'-diol, is reduced by a factor of ~2 in the presence of acrylamide [140]. This contrasts with oxidation by lead(IV) acetate where acrylamide has no effect on the yield. At first glance, these differing results might suggest that lead(IV) reacts in a one-stage two-equivalent process. However, it is possible that two one-equivalent stages are still involved and that the radical is too strongly complexed to react with acrylamide, or is removed much more readily by Pb(IV) or Pb(III) than by acrylamide.

Pinacol (2,3-dimethylbutane-2,3-diol) is oxidised quantitatively to acetone by manganese(III) in perchloric acid media according to

$$2 \text{ Mn(III)} + (CH_3)_2COH \cdot (CH_3)_2COH = 2 \text{ Mn(II)} + 2 (CH_3)_2CO + 2 H^+$$

and a simple second-order rate law applies [141]. The shape of stopped-flow oscilloscope traces indicates two transient complexes and this result, together with the acid dependence of the rate, has led Wells and Barnes to propose the scheme

$$Mn^{3+} + H_2O \rightleftharpoons MnOH^{2+} + H^+$$
$$(CH_3)_2COH \cdot (CH_3)_2COH + H^+ \rightleftharpoons (CH_3)_2COH \cdot (CH_3)_2C\overset{+}{O}H_2$$
$$Mn^{3+} + (CH_3)_2COH \cdot (CH_3)_2COH \rightleftharpoons Mn^{3+} \cdot (CH_3)_2COH \cdot (CH_3)_2COH$$
$$MnOH^{2+} + (CH_3)_2COH \cdot (CH_3)_2COH \rightleftharpoons Mn^{3+} \cdot (CH_3)_2COH \cdot (CH_3)_2CO^- + H_2O$$
$$Mn^{3+} \cdot (CH_3)_2COH \cdot (CH_3)_2COH \rightleftharpoons Mn^{3+} \cdot (CH_3)_2COH \cdot (CH_3)_2CO^- + H^+$$
$$Mn^{3+} \cdot (CH_3)_2COH \cdot (CH_3)_2COH \longrightarrow Mn(II) + (CH_3)_2CO + (CH_3)_2\overset{\cdot}{C}OH + H^+ \quad \text{slow}$$
$$Mn^{3+} \cdot (CH_3)_2COH \cdot (CH_3)_2CO^- \longrightarrow Mn(II) + (CH_3)_2CO + (CH_3)_2\overset{\cdot}{C}OH \quad \text{slow}$$
$$Mn(III) + (CH_3)_2\overset{\cdot}{C}OH \longrightarrow Mn(II) + (CH_3)_2CO + H^+$$

Cerium(IV) in perchlorate media oxidises pinacol in a comparable way to manganese(III) with the intermediate complexes, $Ce^{4+} \cdot (CH_3)_2COH \cdot (CH_3)_2COH$ and $Ce^{4+} \cdot (CH_3)_2COH \cdot (CH_3)_2CO^-$, taking part in the rate-controlling stages [142]

$$Ce^{4+} \cdot (CH_3)_2COH \cdot (CH_3)_2COH \xrightarrow{k_1} Ce(III) + (CH_3)_2\overset{\cdot}{C}OH + (CH_3)_2CO + H^+$$

and

$$Ce^{4+} \cdot (CH_3)_2 COH \cdot (CH_3)_2 CO^- \xrightarrow{k_2} Ce(III) + (CH_3)_2 CO + (CH_3)_2 \dot{C} OH$$

A similarity in the values of k_1 and k_2 (at $27°C$, 0.70 s^{-1} and 0.61 s^{-1}, respectively) shows that the two reactions are closely competitive.

The mechanism proposed by Littler and Waters [143] for the oxidation of pinacol by vanadium(V) is

$$VO_2^+ + (CH_3)_2 COH \cdot (CH_3)_2 COH \rightleftharpoons VO_2^+ \cdot (CH_3)_2 COH \cdot (CH_3)_2 COH$$

$$VO_2^+ \cdot (CH_3)_2 COH \cdot (CH_3)_2 COH \longrightarrow (CH_3)_2 CO + (CH_3)_2 \dot{C} OH + H^+ + VO_2 \quad \text{slow}$$

$$VO_2^+ + (CH_3)_2 \dot{C} OH \longrightarrow (CH_3)_2 CO + H^+ + VO_2$$

$$VO_2 + H_2 O \longrightarrow VO^{2+} + 2 OH^-$$

where vanadium(IV) is produced first as VO_2 and then as VO^{2+}. A dependence on hydrogen ion concentration arises from the formation of $V(OH)_3^{2+}$.

6. OXIDATION OF ALDEHYDES

Both enolisable and non-enolisable aldehydes are attacked by one-equivalent oxidants. The latter group behave like alcohols because of hydrate formation.

The inverse acid dependence found for the reaction between cobalt(III) perchlorate and formaldehyde is interpreted by Hargreaves and Sutcliffe [1] as arising from the protonation of hydrated formaldehyde rather than from the involvement of $CoOH^{2+}$ since the high acidities adopted in their study preclude the formation of appreciable concentrations of hydrolysed species. Accordingly the mechanism suggested is

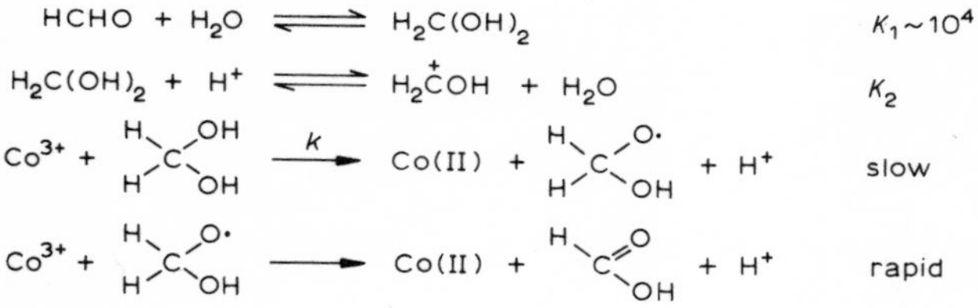

$$\text{HCHO} + \text{H}_2\text{O} \rightleftharpoons \text{H}_2\text{C(OH)}_2 \qquad\qquad K_1 \sim 10^4$$

$$\text{H}_2\text{C(OH)}_2 + \text{H}^+ \rightleftharpoons \text{H}_2\overset{+}{\text{C}}\text{OH} + \text{H}_2\text{O} \qquad\qquad K_2$$

On this basis

$$-\frac{d[\text{Co(III)}]_t}{dt} = k\,[\text{Co}^{3+}][\text{H}_2\text{C(OH)}_2]$$

$$= k_{\text{obs}}\,[\text{Co(III)}]_t\,[\text{formaldehyde}]_t$$

where the subscript t denotes the total concentrations given by

$$[\text{Co(III)}]_t = [\text{Co}^{3+}]$$

and

$$[\text{formaldehyde}]_t = [\text{H}_2\text{C(OH)}_2] + [\text{H}_2\overset{+}{\text{C}}\text{OH}]$$

Thus

$$k_{\text{obs}}[\text{Co}^{3+}]\big([\text{H}_2\text{C(OH)}_2] + [\text{H}_2\overset{+}{\text{C}}\text{OH}]\big) = k\,[\text{Co}^{3+}][\text{H}_2\text{C(OH)}_2]$$

and

$$\frac{1}{k_{\text{obs}}} = \frac{[\text{H}_2\text{C(OH)}_2] + [\text{H}_2\overset{+}{\text{C}}\text{OH}]}{k\,[\text{H}_2\text{C(OH)}_2]}$$

Furthermore

$$\frac{1}{k_{\text{obs}}} = \frac{[\text{H}_2\text{C(OH)}_2] + K_2[\text{H}_2\text{C(OH)}_2][\text{H}^+]}{k\,[\text{H}_2\text{C(OH)}_2]}$$

or

$$\frac{1}{k_{obs}} = \frac{1}{k} + \frac{K_2[H^+]}{k}$$

This relationship is in keeping with the observed acid dependence. Evaluation of the intercept ($= 1/k$) and slope ($= K_2/k$) of the appropriate $1/k_{obs}$ versus $[H^+]$ plot allows values for k and K_2 to be obtained of $20 \text{ l mol}^{-1} \text{ s}^{-1}$ and 0.7 l mol^{-1}, respectively, at $22°C$. Temperature dependences of k and K_2 permit the associated activation energy and enthalpy change to be calculated. In sulphuric acid, the above rate-controlling step is replaced by a slower one between a sulphate complex and an aquated formaldehyde molecule.

$$Co(SO_4)_2^- + H^+ \rightleftharpoons HCo(SO_4)_2$$

$$Co(SO_4)_2^- + H_2C(OH)_2 \xrightarrow{k'} Co(II) + H_2C(OH)O\cdot + H^+ \qquad \text{slow}$$

$$Co(III) + H_2C(OH)O\cdot \longrightarrow Co(II) + HCOOH + H^+ \qquad \text{rapid}$$

where $k' = 7 \text{ l mol}^{-1} \text{ s}^{-1}$ at $22°C$.

In perchlorate media, the rate of oxidation of formaldehyde to formic acid by cerium(IV) is first-order in both substrate and oxidant, the second-order rate constant showing a dependence on acidity of the type $1/k_{obs} = a + (b/[H^+])$ where a and b are constants [144]. No evidence is provided for complex formation at low aldehyde concentrations. Colour changes do occur at higher concentrations but the reaction is then too rapid to measure by conventional spectrophotometry. Hydrolysis of the oxidant is considered important and a likely sequence of steps is

$$Ce^{4+} + H_2O \rightleftharpoons CeOH^{3+} + H^+ \qquad\qquad K$$

$$H_2C(OH)_2 + H^+ \rightleftharpoons H_2\overset{+}{C}OH + H_2O \qquad\qquad K_2$$

which leads to a rate law having the full form

$$\frac{1}{k_{obs}} = \frac{1 + KK_2}{k} + \frac{K}{k[H^+]} + \frac{K_2[H^+]}{k}$$

The last term can be neglected and plots of the reciprocal of the observed rate constant against $1/[H^+]$ are linear. Alternatively, radicals may result from carbon—hydrogen rather than oxygen—hydrogen fission as depicted by

By way of contrast, the acid dependence in sulphuric acid media is $k_{obs} \propto [H_2SO_4]^3$, a relationship which can be rationalised in terms of the existence of a number of Ce(IV)—sulphate equilibria involving the species $Ce(SO_4)_4^{4-}$, $HCe(SO_4)_4^{3-}$, $H_2Ce(SO_4)_4^{2-}$, $H_3Ce(SO_4)_4^{-}$, and $H_4Ce(SO_4)_4$, the latter taking part in the rate-determining stage

$$H_4Ce(SO_4)_4 + H_2C(OH)_2 \longrightarrow Ce(III) + H_2C(OH)O\cdot + H^+$$

A particularly interesting feature of the oxidation of formaldehyde (and also formic acid) by manganese(III) sulphate in sulphuric acid is the retardation brought about by manganese(II) ions [145]. This observation can be interpreted in the terms of the equilibrium

$$2\,Mn(III) \rightleftharpoons Mn(IV) + Mn(II)$$

On the assumption that Mn(IV) is the active oxidising species, the rate is given by

$$-\frac{d[\text{Mn(III)}]}{dt} = \frac{k\,[\text{aldehyde}]\,[\text{Mn(III)}]^2}{[\text{Mn(II)}]}$$

Formaldehyde is oxidised by V(V) perchlorate according to the rate equation [146]

$$-\frac{d[\text{V(V)}]}{dt} = k_1[\text{substrate}]\left\{[\text{V(V)}][\text{H}^+] + k_2[\text{V(V)}]^2\,[\text{H}^+]^2\right\}$$

The second term could arise from the reaction of a V(V)— substrate complex with more V(V). A primary kinetic isotope effect (of ~4.5) indicates that C—H bond fission is important.

The ferricyanide oxidation of formaldehyde is base-catalysed and the kinetics are in keeping with the scheme [147]

$$
\begin{aligned}
&\text{CH}_2(\text{OH})_2 + \text{OH}^- \underset{k_{-1}}{\overset{k_1}{\rightleftharpoons}} \text{CH}_2(\text{O}^-)(\text{OH}) + \text{H}_2\text{O}\\[4pt]
&\text{Fe(CN)}_6^{3-} + \text{CH}_2(\text{O}^-)(\text{OH}) \overset{k_2}{\longrightarrow} \text{Fe(CN)}_6^{4-} + \text{CH}_2(\text{O}\cdot)(\text{OH})\\[4pt]
&\text{Fe(CN)}_6^{3-} + \text{CH}_2(\text{O}\cdot)(\text{OH}) \longrightarrow \text{Fe(CN)}_6^{4-} + \text{HCOOH} + \text{H}^+ \qquad \text{rapid}
\end{aligned}
$$

and the associated rate law

$$-\frac{d[\text{Fe(CN)}_6^{3-}]}{dt} = \frac{2\,k_1 k_2\,[\text{Fe(CN)}_6^{3-}]\,[\text{aldehyde}]\,[\text{OH}^-]}{k_{-1}[\text{H}_2\text{O}] + k_2[\text{Fe(CN)}_6^{3-}]}$$

Formaldehyde reduces alkaline solutions of Cu(II), not to Cu(I) oxide, but to metallic copper. Shanker and Singh [148] (using alkaline glycerol solutions) have found a rate law, first-order in both Cu(II) and aldehyde concentrations but second-order in OH$^-$ concentration in accordance with the following hydride transfer mechanism.

Isobutyraldehyde is oxidised (to acetone and formic acid) by vanadium(V) in perchloric acid solution at a rate which is independent of the concentration of the oxidant [149]. This suggests that the rate-determining stage is enolisation of the aldehyde as is known to be the case in halogen substitution. Indeed, the rates of substitution of the aldehyde by iodine and bromine are shown to be identical with the rate of oxidation by V(V) (Fig. 3). By way of contrast, propionaldehyde and n-butyraldehyde are oxidised (to acetaldehyde and formic acid) more slowly than they enolise and the rate law is

$$-\frac{d[V(V)]}{dt} = \frac{k[V(V)][\text{aldehyde}]}{1 + K[\text{aldehyde}]}$$

which fits the general scheme

$$V(V) + \text{aldehyde} \xrightarrow{K} \text{complex} \xrightarrow{k} \text{products}$$

The acid dependence of the reactions implies that the complexes are protonated.

Although a mild oxidising agent, copper(II) is capable of

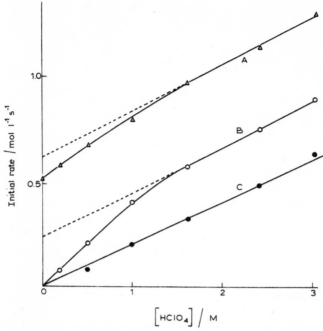

Fig. 3. Acid dependence of rate of reaction of isobutyraldehyde with A, 0.05 N bromine; B, 0.05 N iodine; and C, 0.05 N vanadium(V); [aldehyde] = 0.22 M, 25°C, ionic strength = 3.2 M. (From Jones and Waters [149], by courtesy of The Chemical Society.)

oxidising enolisable aldehydes to their corresponding carboxylate anions as evidenced by the well-known Fehling and Benedict tests for aliphatic aldehydes (which use Cu(II) in alkaline tartrate and citrate media, respectively). It appears that the first stage in the aldehyde oxidation is the removal of an α proton, generating an enolate anion which is subsequently oxidised by Cu(II).

$$RCH_2CHO + OH^- \;\rightleftharpoons\; \left[\underset{O}{R\bar{C}HCH} \;\longleftrightarrow\; \underset{O^-}{RCH=CH} \right] + H_2O$$

$$\downarrow Cu(II)$$

$$\left[\underset{O}{R\dot{C}HCH} \;\longleftrightarrow\; \underset{O\cdot}{RCH=CH} \right] + Cu(I)$$

$$\downarrow Cu(II)$$

$$RCH_2COO^-$$

Side products arise from the oxidation of $R\dot{C}HCHO$ to the α-hydroxyaldehyde which in turn is oxidised further. The copper(I) product appears as the oxide.

Chlorination of isobutyraldehyde by Cu(II) chloride in aqueous acetone has a rate law given by [150]

$$-\frac{d[Cu(II)]}{dt} = k_1[CuCl_2][RCHO] + k_2[H^+][CuCl_2][RCHO]$$

and is accelerated by chloride ion, reaching a limit at a $Cl^-/CuCl_2$ ratio of 2 which suggests that $CuCl_4^{2-}$ is involved in the rate-determining step. Similar results were obtained for the chlorination of n-butyraldehyde in dimethylformamide [151], a simple second-order rate law applying

$$\frac{d[Cu(I)]}{dt} = k[CuCl_2][RCHO]$$

The participation of active chlorine resulting from the reaction

$$2\,CuCl_2 \;\rightleftharpoons\; 2\,CuCl + Cl_2$$

is hardly likely on the grounds of the moderate temperature used (30—80°C) and Nigh [152] has proposed, instead, the following general scheme for these types of oxidative halogenation.

$$Cu(II) + RCH_2\underset{\underset{O}{\|}}{C}R' \rightleftharpoons RCH_2\underset{\underset{O}{\|}}{\underset{|}{C}}R' \xrightarrow{H^+} RCH=\underset{\underset{O}{|}}{\underset{|}{C}}R' + H^+$$
$$\quad\quad\quad\quad\quad\quad\quad\quad\quad Cu(II)\quad\quad\quad\quad\quad Cu(II)$$

$$RCH=\underset{\underset{O}{|}}{\underset{|}{C}}R' + Cu(II) \longrightarrow R\underset{\underset{O}{\|}}{C}H\overset{\overset{Cl}{|}}{C}R' + 2Cu(I)$$
$$\quad\; Cu(II)$$

where Cu(II) is $CuCl_3^-$ or $CuCl_4^{2-}$. Simple unassisted enolisation can be ruled out in that the rate of chlorination of isobutyraldehyde is greater than the rate of direct enolisation.

Cobalt(III) is sufficiently powerful to oxidise aromatic aldehydes. Because of the lack of solubility of the aldehydes in water, Cooper and Waters [153] employed a mixed solvent of methyl cyanide and water in their kinetic study of the oxidation of *m*- and *p*-nitrobenzaldehydes by cobalt(III) perchlorate. First-order in both Co(III) and substrate, the reaction rate is inversely proportional to hydrogen ion concentration. By using m-$NO_2.C_6H_4.CDO$, a primary kinetic isotope effect (of 2.3 at 10°C) was demonstrated, a result indicative of C—H bond fission in the rate-determining step

$$Co(III) + ArCHO \longrightarrow Co(II) + ArCO\cdot + H^+ \quad\quad slow$$

Radicals are then removed by reaction with oxidant

$$Co(III) + ArCO\cdot + H_2O \longrightarrow Co(II) + ArCOOH + H^+ \quad\quad rapid$$

The slow stage may involve either the removal of a hydrogen atom by the OH group of the hydroxo Co(III) ion in an outer-sphere process

$$Ar\underset{\underset{O}{\diagdown}}{\overset{\overset{H}{\diagup}}{C}} + HO-Co(OH_2)_5^{2+} \longrightarrow ArCO\cdot + H_2O + Co(II)_{aq}$$

or the formation of an aldehyde—cobalt(III) complex followed by the inner-sphere process

$$ArC\overset{O}{\underset{H}{\diagdown}}\underset{OH}{\diagup}Co^{III}(OH_2)_4 \longrightarrow ArCO\cdot + Co(II)_{aq} + H_2O$$

The oxidation of benzaldehyde by cobalt(III) acetate in acetic acid—sulphuric acid mixtures, investigated as part of a study of the autoxidation of benzaldehyde in the presence of cobalt(II) salts [154], appears to take a course similar to that above.

Wiberg and Ford [155] have made a thorough examination of the kinetics of oxidation of benzaldehyde by cerium(IV) perchlorate. The medium chosen was 85% aqueous acetic acid. Based on the expected 2:1 stoichiometry, the yield of benzoic acid is 89%, a little benzil being formed as by-product. A complication is that oxygen strongly accelerates the reaction and careful degassing of the reactant solutions (before mixing) was found necessary. Further complications are that in aqueous acetic acid, benzaldehyde exists in both monomeric and trimeric forms [156] and that Ce(IV) is slowly reduced by the solvent. Optical evidence indicates that two complexes (C_1 and C_2) are formed between oxidant and substrate

$$Ce^{4+} + R'CHO \rightleftharpoons C_1$$

$$C_1 + R'CHO \rightleftharpoons C_2$$

The oxidation steps, depicted as

$$C_1 + H^+ \longrightarrow R'CO\cdot + Ce(III)$$

and

$$C_2 + 2H^+ \longrightarrow R'CO\cdot + Ce(III) + R'CHO$$

are followed by

$$R'CO \cdot + Ce(IV) \longrightarrow R'COOH + Ce(III)$$

Kinetic isotope effects indicate that the oxidation steps involve C—H bond cleavage and the results of induced polymerisation experiments confirm that the benzoyl radical is the active species.

7. OXIDATION OF KETONES

Numerous kinetic studies have appeared on the oxidation of ketones. In some of these, there is evidence to suggest that primary attack takes place on an enol form of the ketone; in others, direct reaction with the ketone is indicated.

The kinetics of oxidation of aliphatic ketones by cerium(IV) sulphate were first studied by Shorter and Hinshelwood [157a, 157b]. These authors suggested that enolisation was basic to the mechanism of oxidation. A further study by Shorter [158] of the oxidation of acetone by ammonium ceric nitrate in aqueous nitric acid solution substantiated the previous suggestions, although an attempt to confirm the presence of an enolic intermediate (by adding bromine to remove the enol as soon as it is formed) was only partially successful (because of the bromine-catalysed oxidation of water). The products of the oxidation are formic and acetic acids, six equivalents of Ce(IV) being consumed per mole of acetone oxidised. Likely intermediates are acetol (CH_3COCH_2OH) and methylglyoxal (CH_3COCHO). However, reaction with cerium(IV) sulphate in sulphuric acid (at 70°C) results in the consumption of 8.6 equivalents of Ce(IV), both methyl groups being attacked.

Oxidation of cyclohexanone by cerium(IV) sulphate and manganese(III) sulphate is much faster than enolisation and the rate is first-order in oxidant concentration. It appears that the oxidant attacks the ketone directly, an α C—H bond being broken in the rate-determining step. Littler [159] points out that attack on either ketone or enol would produce the same (mesomeric) radical, viz.

but the much lower concentration of the enolic form ensures that reaction with the ketone predominates. These conclusions are in direct contrast to those reached for the oxidation of the same ketone by the multi-equivalent oxidants, I_2, Br_2, mercury(II), thallium(III), and manganese(VII), where the rate of oxidation is zero-order in oxidant concentration and enolisation is rate-determining [160]. Significantly, these oxidants, unlike the one-equivalent oxidants, react readily with olefins and the mechanism of reaction with the enol form may be essentially attack on an isolated double bond by way of complex formation (p. 143).

Hoare and Waters [161] have examined the kinetics of oxidation of diethyl ketone by cobalt(III) perchlorate and sulphate and have suggested that C—C bond fission occurs at an early stage in the reaction sequence, possibly by an electron transfer to give O—$\overset{+}{C}(C_2H_5)_2$ which then breaks down to $O=\overset{\cdot}{C}(C_2H_5)$ and an ethyl radical. The same mechanism would seem to hold for the cobalt(III) oxidation of cyclohexanone under acid conditions [159].

The tris-o-phenanthroline iron(III) complex [162] and the hexachloroiridate(IV) [163] anion oxidise cyclohexanone by outer-sphere processes involving electron transfer to the enol form. In the latter oxidation, the initial products are the hexachloroiridate(III) ion and a 2-oxocyclohexyl radical. The radical is oxidised further by an inner-sphere (ligand transfer) process to give 2-chlorocyclohexanone and aquopentachloro-iridate(III).

Oxygen and acrylonitrile compete with iridium(IV) for the radical intermediate. As would be expected on this mechanism, the rate of oxidation (in perchloric acid) is twice the rate of enolisation.

Copper(II) chloride is readily reduced to copper(I) chloride by carbonyl compounds which in turn are oxidised to the corresponding chloro compounds. With acetone the reaction is

$$2\,CuCl_2 + CH_3COCH_3 = 2\,CuCl + ClCH_2COCH_3 + HCl$$

In aqueous chloride solution, the reaction is retarded by copper(I), a limit being attained when the amount of Cu(I) added is approximately half that of Cu(II). Initial rate measurements indicate the kinetics to be half-order in total Cu(II) concentration and first-order in chloride ion. Kochi [164] explains the retarding influence of Cu(I) chloride as arising from the formation of an unreactive Cu(II)—Cu(I) complex

$$CuCl + CuCl_2 \rightleftharpoons Cu_2Cl_3$$

and the observed half-order dependence on Cu(II) as being brought about by the formation of an unreactive dimeric Cu(II) species

$$2\,CuCl_2 \; \rightleftharpoons \; Cu_2Cl_4$$

These equilibria are then followed by the slower reaction

$$CuCl_2 + CH_3COCH_3 \longrightarrow Cu + ClCH_2COCH_3 + HCl$$

where Cu(0) is in equilibrium with $CuCl_2$

$$Cu + CuCl_2 \rightleftharpoons 2\,CuCl$$

However, it seems possible that $CuCl_3^-$ could well be involved [165]

$$CuCl_2 + Cl^- \rightleftharpoons CuCl_3^-$$

$$CuCl_3^- + CH_3COCH_3 \longrightarrow Cu + ClCH_2COCH_3 + HCl + Cl^-$$

thereby explaining the first-order dependence of the rate on chloride concentration.

8. OXIDATION OF CARBOXYLIC ACIDS

The first part of this section deals with the oxidation of formic acid. Following this, an account is given of the oxidation of other aliphatic carboxylic acids. Dicarboxylic acids also react with one-equivalent oxidants and are considered last.

Wells and Whatley [166] have examined the oxidation of formic acid by manganese(III) in perchloric acid

$$2\,Mn(III) + HCOOH = 2\,Mn(II) + CO_2 + 2H^+$$

With formic acid in excess, the rate of decay of Mn(III) is
first-order for both aerobic and anaerobic conditions. The lack
of effect of Mn(II) rules out the involvement of Mn(IV) arising
from the disproportionation

$$2Mn(III) \rightleftharpoons Mn(IV) + Mn(II)$$

With oxygen excluded, reaction with added acrylamide results
in rapid precipitation of polymer, thus indicating the presence
of radicals. Although formic acid has little effect on the
Mn(III) absorption peak at 470 nm, marked increase in light
absorption occurs in the u.v. region from 260—300 nm
(analogous effects are noted for alcohols and glycols, see pp. 48
and 55). Plots of the observed first-order rate constant (k_{obs})
versus substrate concentration are non-linear, tending towards
a limiting value at high acid concentrations. A linear relationship exists, however, between $1/k_{obs}$ and $1/[HCOOH]$, both
intercept and slope varying with acidity at constant temperature. The general observations and kinetic data fit a mechanism
in which oxidant—substrate complexes decay to give ·COOH
radicals which are then oxidised further by Mn(III).

$$Mn^{3+} + HCOOH \rightleftharpoons Mn^{3+} \cdot HCOOH \qquad K_1$$

$$Mn^{3+} \cdot HCOOH \rightleftharpoons Mn^{3+} \cdot HCOO^- + H^+ \qquad K_2$$

$$Mn^{3+} \cdot HCOOH \xrightarrow{k_1} Mn(II) + \cdot COOH + H^+ \qquad slow$$

$$Mn^{3+} \cdot HCOO^- \xrightarrow{k_2} Mn(II) + \cdot COOH \qquad slow$$

$$Mn(III) + \cdot COOH \longrightarrow Mn(II) + CO_2 + H^+ \qquad rapid$$

Thus the rate law, in terms of the disappearance of total
Mn(III), becomes

$$-\frac{d[Mn(III)]_t}{dt} = 2k_1[Mn^{3+} \cdot HCOOH] + 2k_2[Mn^{3+} \cdot HCOO^-]$$

Taking into account the hydrolysis equilibrium

$$Mn^{3+} + H_2O \rightleftharpoons MnOH^{2+} + H^+ \qquad K_h$$

it follows that

$$\frac{1}{k_{obs}} = \frac{(1 + K_h/[H^+])}{2K_1(k_1 + k_2K_2/[H^+])[HCOOH]} + \frac{(1 + K_2/[H^+])}{2(k_1 + k_2K_2/[H^+])}$$

Intercepts and slopes of plots of $1/k_{obs}$ versus $1/[HCOOH]$ thus vary with acidity. Although the above mechanism holds at temperatures above 40°C, detailed examination of the acidity dependence of the slopes reveals that at lower temperatures a further (hydrolysed) Mn(III) complex participates

$$Mn^{3+}.HCOOH + H_2O \rightleftharpoons (MnOH)^{2+}.HCOO^- + 2 H^+$$

$$(MnOH)^{2+}.HCOO^- \longrightarrow Mn(II) + \cdot COO^- + H_2O$$

Kemp and Waters [145] note that the kinetics of oxidation of formic acid by Mn(III) sulphate in sulphate media are described by a rate law having a $[Mn(III)]^3$ term rather than a $[Mn(II)]^2$ term and suggest that the slow stage is either the attack of Mn(IV) on a Mn(III)—formate complex, or attack of Mn(III) on a Mn(IV)—formate complex. Formic acid is oxidised very slowly by cerium(IV) in perchlorate media [167]. Kinetic and spectral evidence indicates the existence of complexes $Ce^{4+}.HCOOH$ and $Ce^{4+}.HCOO^-$ and the mechanism has a general resemblance to that for the formic acid—manganese(III) system.

From kinetic and products studies, Clifford and Waters [168] have shown that the oxidation of aliphatic carboxylic acids by cobalt(III) perchlorate involves the rapid reversible formation of a Co(III) complex

$$Co(H_2O)_6^{3+} + RCOOH \rightleftharpoons RCOOCo(H_2O)_5^{2+} + H^+ + H_2O \qquad K_1$$

which then undergoes slow fragmentation to liberate free alkyl radicals

$$RCOOCo(H_2O)_5^{2+} \xrightarrow{k_2} R\cdot + CO_2 + Co(II) + 5 H_2O$$

These then participate in further rapid reactions, e.g.

$$R\cdot + Co(III) \longrightarrow R^+ + Co(II)$$
$$R^+ + H_2O \longrightarrow ROH + H^+$$

Michaelis—Menten kinetics are followed. Since

$$\left[RCOOCo(H_2O)_5^{2+}\right] = \frac{\left[Co(III)\right]_t K_1 \left[RCOOH\right]}{\left[H^+\right] + K_1 \left[RCOOH\right]}$$

then

$$-\frac{d\left[Co(III)\right]_t}{dt} = k_{obs}\left[Co(III)\right]_t = \frac{k_2 \left[Co(III)\right]_t K_1 \left[RCOOH\right]}{\left[H^+\right] + K_1 \left[RCOOH\right]}$$

and

$$\frac{1}{k_{obs}} = \frac{1}{k_2} + \frac{\left[H^+\right]}{k_2 K_1 \left[RCOOH\right]}$$

Plots of $1/k_{obs}$ versus $1/[RCOOH]$ are linear with intercepts of $1/k_2$ and slopes of $[H^+]/k_2K_1$. The ease of oxidation of aliphatic acids follows the order $CH_3COOH < CH_3CH_2COOH < (CH_3)_2CHCOOH < (CH_3)_3CCOOH$, the energy of activation decreasing by approximately 17 kJ mol^{-1} for successive replacements of the α hydrogen atoms by methyl groups. This result could hardly arise from a two-stage decomposition of the type

$$RCOOCo^{III} \longrightarrow RCOO\cdot + Co(II) \qquad slow$$
$$RCOO\cdot \longrightarrow R\cdot + CO_2 \qquad rapid$$

because the inductive effect of the distant alkyl group on the fission of a Co—O bond would be expected to be slight. More likely is a concerted fragmentation with concurrent formation of an alkyl radical and carbon dioxide. Radicals generated in this way have been shown to bring about the oxidation of toluene by hydrogen transfer [169]

$$R\cdot \; + \; C_6H_5CH_3 \; \longrightarrow \; RH \; + \; C_6H_5CH_2\cdot$$

Furthermore, alkyl radicals formed during the oxidation of acids of the type $C_6H_5(CH_2)_n COOH$ can be trapped by bromoform

$$R\cdot \; + \; Br_3CH \; \longrightarrow \; RBr \; + \; \cdot CHBr_2$$

and estimated as bromides [170].

The kinetics of oxidation of acetic acid by cerium(IV) in aqueous perchloric acid have been studied [171] by following the rate of disappearance of Ce(IV) both volumetrically and by measurement of the gases evolved (methane and carbon dioxide). In the temperature range 50—60°C, the reaction was found to be first-order in cerium(IV) and approximately first-order in acetic acid, the rate increasing with hydrogen ion concentration (with a linear plot of $1/k_{obs}$ versus $1/[H^+]$ where k_{obs} is the observed first-order rate constant). A linear relation exists also between $1/k_{obs}$ and $1/[CH_3COOH]$. These Michaelis—Menten kinetics point to the existence of initial complex formation between Ce(IV) and acetic acid. Supporting spectrophotometric evidence comes from the observation that the presence of acetic acid gives rise to an additional absorption peak (at 286 nm) in the spectrum of Ce(IV) perchlorate (which has maxima at 240 and 251 nm). The following inner-sphere mechanism accounts for the observed results.

$$Ce(H_2O)_7(OH)^{3+} + H^+ \rightleftharpoons Ce(H_2O)_8^{4+}$$

$$Ce(H_2O)_8^{4+} + CH_3COOH \rightleftharpoons Ce(H_2O)_7(CH_3COOH)^{4+} + H_2O$$

$$Ce(H_2O)_7(CH_3COOH)^{4+} \rightleftharpoons Ce(H_2O)_7(CH_3COO)^{3+} + H^+$$

$$Ce(H_2O)_7(CH_3COO)^{3+} \rightleftharpoons Ce^{III}(H_2O)_6 + CH_3COO\cdot + H_2O$$

$$CH_3COO\cdot \longrightarrow CH_3\dot{} + CO_2$$

As expected, alkyl-substituted acids react more quickly than acetic acid whereas halogen substitution decreases the rate; presumably, groups which increase the electron density at the reaction centre facilitate electron transfer. Lande and Kochi [172] have studied the formation and oxidation of alkyl radicals during the thermal decarboxylation of carboxylic acids by cobalt(III). Manganese(III) oxidises acetic acid at

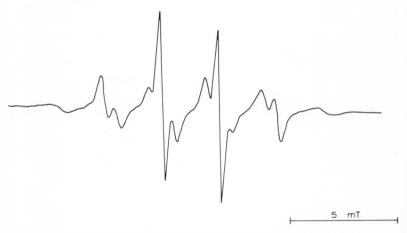

5 mT

Fig. 4. E.s.r. spectrum of ethyl radical generated during photolysis of an aqueous acidic solution of Ce(IV) and propionic acid at 77 K. (From Greatorex and Kemp [174], by courtesy of The Chemical Society.)

100°C to give, principally, acetoxyacetic acid and carbon dioxide [173]. The low yield of methane indicates that the $\cdot CH_2COOH$ radical, rather than the $CH_3COO\cdot$ radical, is the primary species (compare the decomposition of Pb(IV) acetate, p.137). Alkyl radicals, generated during the photodecomposition of complexes of cerium(IV) and carboxylic acids in aqueous perchloric acid, have been detected by electron spin resonance [174]. Oxidant and reductant are allowed to interact before the solution is frozen and u.v. irradiation is used to bring about the electron transfer process. The alkyl radicals produced are then trapped and prevented from reacting further with oxidant (Fig. 4).

Isobutyric acid is oxidised by manganese(III) in perchlorate media to give isopropanol, acetone, and carbon dioxide; the yield of acetone is low under anaerobic conditions [175]. Oxygen has no effect on the rate and the reaction is first-order in both Mn(III) and isobutyric acid concentrations, being independent of changes in acidity. No effects occur in the absorption spectrum of Mn(III) on addition of isobutyric acid and hence complex formation appears unimportant. The presence of radicals in the reacting mixture is demonstrated by the rapid polymerisation of acrylonitrile in the absence of oxygen, and Wells and Barnes [175] suggest the following outer-sphere mechanism.

$$Mn(III) + (CH_3)_2CHCOOH \longrightarrow Mn(II) + (CH_3)_2\overset{+}{C}HCOOH$$

$$(CH_3)_2CH\overset{+}{C}OOH \longrightarrow (CH_3)_2\dot{C}H + CO_2 + H^+$$

$$Mn(III) + (CH_3)_2\dot{C}H \longrightarrow (CH_3)_2\overset{+}{C}H + Mn(II)$$

$$(CH_3)_2\overset{+}{C}H + H_2O \longrightarrow (CH_3)_2CHOH + H^+$$

With oxygen present, the following additional steps are introduced.

$$(CH_3)_2\dot{C}H + O_2 \longrightarrow (CH_3)_2CHO_2\cdot$$

$$2 \ (CH_3)_2CHO_2\cdot \longrightarrow 2 \ (CH_3)_2CHO\cdot + O_2$$

$$Mn(III) + (CH_3)_2CHO\cdot \longrightarrow (CH_3)_2CO + Mn(II) + H^+$$

and oxygen competes with manganese(III) for the radical intermediate $(CH_3)_2\dot{C}H$.

Davies and Watkins [176a] have examined in detail the kinetics of the oxidation of oxalic acid by cobalt(III) in acid perchlorate solutions. The stoichiometry corresponds accurately to

$$2 \ Co(III) + H_2C_2O_4 = 2 \ Co(II) + 2 \ CO_2 + 2 \ H^+$$

The simple empirical rate law

$$-\frac{d[Co(III)]}{dt} = k_{obs}[Co(III)][H_2C_2O_4]_t$$

was obtained from spectrophotometric observations on the rate of disappearance of Co(III) at 250—270 nm with oxalic acid in excess. The acid dependence is given by

$$k_{obs} = \frac{A + B/[H^+] + C/[H^+]^2}{1 + K_{1a}/[H^+]}$$

where A, B, and C are constants, and K_{1a} is defined in terms of the equilibrium

$$H_2C_2O_4 \rightleftharpoons H^+ + HC_2O_4^- \qquad K_{1a}$$

which occurs along with

$$HC_2O_4^- \rightleftharpoons H^+ + C_2O_4^{2-} \qquad K_{2a}$$

The following sequence of steps is possible.

$$Co^{3+} + H_2O \rightleftharpoons CoOH^{2+} + H^+ \qquad K_h$$

$$Co^{3+} + H_2C_2O_4 \xrightarrow{k_1} Co(II) + Ox\cdot$$

$$CoOH^{2+} + H_2C_2O_4 \xrightarrow{k_2} Co(II) + Ox\cdot$$

$$Co^{3+} + HC_2O_4^- \xrightarrow{k_3} Co(II) + Ox\cdot$$

$$CoOH^{2+} + HC_2O_4^- \xrightarrow{k_4} Co(II) + Ox\cdot$$

$$Co^{3+} + C_2O_4^{2-} \xrightarrow{k_5} Co(II) + Ox\cdot$$

$$CoOH^{2+} + C_2O_4^{2-} \xrightarrow{k_6} Co(II) + Ox\cdot$$

$$Co(III) + Ox\cdot \longrightarrow Co(II) + CO_2 \qquad rapid$$

Although no evidence is evinced for the formation of complexes, mono-oxalato species are likely to be formed as intermediates. The derived rate law is complex

$$k_{obs} = 2\left[\frac{k_1 + (k_2 K_h + k_3 K_{1a})/[H^+] + K_{1a}(k_4 K_h + k_5 K_{2a})/[H^+]^2 + k_6 K_h K_{1a} K_{2a}/[H^+]^3}{(1 + K_h/[H^+])(1 + K_{1a}/[H^+] + K_{1a} K_{2a}/[H^+]^2)} \right]$$

but reduces to the form of the empirical expression if (a) $K_h/[H^+] \ll 1$, (b) step k_6 is neglected, and (c) $K_{1a} K_{2a}/[H^+]^2 \ll 1 + K_{1a}/[H^+]$. Measurements on K_h, K_{1a}, and K_{2a} show that assumptions (a) and (c) are realistic. The constants A, B, and C are then identified as $A = 2k_1$, $B = 2(k_2 K_h + k_3 K_{1a})$, and $C = 2K_{1a}(k_4 K_h + k_5 K_{2a})$. From previous work, $k_2 K_h \gg k_3 K_{1a}$ and $k_4 K_h \gg k_5 K_{2a}$, and the expressions for B and C simplify to $B = 2k_2 K_h$ and $C = 2k_4 K_{1a} K_h$.

On mixing relatively high concentrations of cerium(IV) and oxalic acid in sulphuric acid media, an intense coloration is rapidly formed which then gradually fades. This is attributed to the formation of an intermediate complex and, as expected, extrapolation of first-order plots of log [Ce(IV)] versus time to zero time gives values of [Ce(IV)] considerably lower than those present initially [176b]. Addition of bisulphate ions, besides causing a decrease in the rate of reaction, decreases the concentration of the intermediate complex although the first-order rate of decomposition of the latter is unaffected. On the basis of the mechanism

$$H_2C_2O_4 \;\rightleftharpoons\; HC_2O_4^- + H^+ \qquad\qquad K$$

$$Ce(SO_4)_3^{2-} + H^+ \;\xrightleftharpoons[k_{-1}]{k_1}\; Ce(SO_4)_2 + HSO_4^- \qquad\qquad K_1$$

$$Ce(SO_4)_2 + HC_2O_4^- \;\rightleftharpoons\; Ce(SO_4)_2(C_2O_4)^{2-} + H^+ \qquad\qquad K_2$$

$$Ce(SO_4)_2(C_2O_4)^{2-} \;\xrightarrow{k_3}\; Ce(SO_4)_2^- + C_2O_4^{\overset{\text{.}}{=}}$$

$$Ce(SO_4)_3^{2-} + C_2O_4^{\overset{\text{.}}{-}} \;\xrightarrow{k_4}\; Ce(SO_4)_2^- + SO_4^{2-} + 2\,CO_2$$

the rate of disappearance of Ce(IV) is

$$-\frac{d\left[Ce(SO_4)_3^{2-}\right]}{dt} = k_1\left[Ce(SO_4)_3^{2-}\right]\left[H^+\right] - k_{-1}\left[Ce(SO_4)_2\right]\left[HSO_4^-\right] + k_4\left[Ce(SO_4)_3\right]\left[C_2O\right.$$

Application of the steady-state assumption to the concentration of the radical reduces this expression to

$$-\frac{d\left[Ce(SO_4)_3^{2-}\right]}{dt} = \left(\frac{K_1 K_2 Kk_3}{[H^+]\left[HSO_4^-\right]}\right)\left[Ce(SO_4)_3^{2-}\right]\left[H_2C_2O_4\right]$$

with the decomposition of the complex as the rate-determining stage. It is clear that the role of added bisulphate is to suppress the formation of the reactive oxalate complex by competing with oxalate for the available cation; phosphate has a similar retarding influence. In non-complexing perchlorate media, the Ce(IV)—oxalic acid reaction is too rapid for conventional spectrophotometric measurements. Inhibition of the reaction can occur at high concentrations of oxalic acid, the rate passing through a maximum and then decreasing as the oxalic acid concentration is increased [177]. This effect is explainable in terms of the formation of less-reactive, higher oxalate complexes. An analytical implication is that in the titration of oxalate with cerium(IV), an induction period is observed during the first stages of the titration.

Kinetically, the reaction between vanadium(V) perchlorate and oxalic acid has some interesting features [178]. The rate of oxidation is acid-dependent, the rate decreasing as the hydrogen ion concentration is increased, achieving a minimum value at $[H^+] \sim 2.5$ M and then increasing. These results suggest that oxalic acid forms with V(V) a series of complexes of varying stability. At low acidities, the reaction appears to be essentially one between a V(V) complex and a further molecule of oxalic acid whereas at high acidities, the reaction becomes an acid-catalysed one between $V(OH)_3^{2+}$ and $H_2C_2O_4$. Oxidations of malonic and oxalic acids by vanadium(V) sulphate are catalysed by manganese(II) ions but retarded by vanadium(IV) [179]. The kinetic results can be interpreted as arising from the slow equilibrium

$$V(V) + Mn(II), L_n \rightleftharpoons V(IV) + Mn(III), L_n \quad \text{slow}$$

where the two Mn oxidation states are complexes containing one mole of substrate each. That the equilibrium is normally displaced markedly to the left is evident from the observation that concentrated solutions of V(V) and Mn(II) generate no detectable amounts of V(IV) or Mn(III) even if held at 60°C for several hours. Electron transfer from the Mn(II) chelate to V(V) is followed by the rapid breakdown of the Mn(III) complex

$$Mn(III), L_n \longrightarrow \text{products} \quad \text{rapid}$$

The rate approaches a limiting value as the amount of added Mn(II) is increased, the limit corresponding to maximum complexing of the cation.

An unusual kinetic study is that between neptunium(VI) and oxalic acid in perchlorate media. Complex formation takes place between Np(VI) and oxalate as is evidenced by a colour change from pink to yellow. The reaction was followed by measurements of the Np(V) concentration at 983 nm and the rate law is

$$\frac{d\left[Np(V)\right]}{dt} = \frac{k\left[Np(VI)\right]\left[oxalate\right]_t}{\left[H^+\right]}$$

Shastri and Amis [180] propose as an initial step

$$NpO_2^{2+} + H_2C_2O_4 + H_2O \rightleftharpoons NpO_2(OH)H_2C_2O_4^+ + H^+ \qquad \text{slow}$$

which is followed by a number of steps summarised by

$$3\ NpO_2^{2+} + NpO_2(OH)H_2C_2O_4^+ \longrightarrow 4\ NpO_2^+ + 2\ CO_2 + 3\ H^+ + \tfrac{1}{2}O_2 \qquad \text{rapid}$$

to give the net reaction

$$4\ NpO_2^{2+} + H_2C_2O_4 + H_2O = 4\ NpO_2^+ + 2\ CO_2 + 4\ H^+ + \tfrac{1}{2}O_2$$

Cerium(IV) in the presence of malonic acid and bromate comprises an oscillating system in that the Ce(IV) is alternately reduced and oxidised according to the overall reactions [181]

$$6\ Ce(IV) + CH_2(COOH)_2 + 2\ H_2O = 6\ Ce(III) + HCOOH + 2\ CO_2 + 6\ H^+$$
$$10\ Ce(III) + 2\ BrO_3^- + 12\ H^+ = 10\ Ce(IV) + Br_2 + 6\ H_2O$$

This behaviour can be demonstrated by the periodic changes in potential of an inert electrode, or visually by means of a

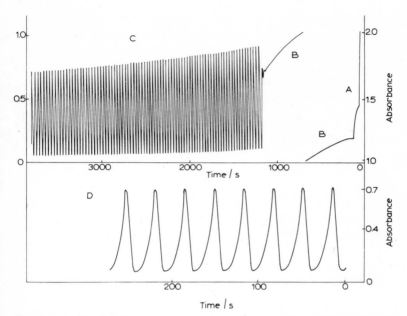

Fig. 5. A trace of absorbance versus time for the oscillating reaction of
1.1×10^{-4} M Ce(IV) sulphate, 0.1 M malonic acid, and 0.03 M potassium
bromate in 3 N H_2SO_4 at 30°C. (From Kasperek and Bruice [182], by
courtesy of The American Chemical Society.)

suitable redox indicator. Kasperek and Bruice [182] have
explored the reaction by making absorbance measurements
on the appearance and disappearance of Ce(IV) at 320 nm as
a function of time. A typical trace is shown as Fig. 5; initially
Ce(IV) decays (region A), an induction period exists (B), and
then abruptly the rapid oscillations begin (C) and continue
for over 500 oscillations. In the lower plot (D) a part of the
oscillatory region has been expanded. Attempts to unravel
the mechanism for this interesting system are hampered by

the complexity (and non-reproducibility) of the Ce(III) + bromate reaction. Furthermore, the sensitivity of the oscillating system to stirring suggests a degree of heterogeneity. Previous explanations appear to be untenable. The phenomenon is not limited to malonic acid; citric, maleic, malic, bromomalonic, and dibromomalonic acids react in a similar periodic manner.

9. OXIDATION OF HYDROXY KETONES AND ACIDS

Differences in the vanadium(V) oxidation of the hydroxy ketones 3-hydroxy-3-methylbutan-2-one, $(CH_3)_2C(OH)COCH_3$, and acetoin, $CH_3CH(OH)COCH_3$, are revealing. The former consumes 2 equivalents of V(V) to give acetone as the product whereas acetoin consumes 4 equivalents to give some biacetyl. Jones and Waters [183] conclude that oxidation of the C-methyl compound

$$2 \ V(\underline{V}) \ + \ (CH_3)_2C(OH)COCH_3 \ \longrightarrow \ 2 \ V(\underline{IV}) \ + \ (CH_3)_2CO \ + \ CH_3COOH$$

resembles the oxidation of pinacol (p. 56) and that C—C fission takes place

$$VO_2^+ \cdot (CH_3)_2C(OH)COCH_3 \ \longrightarrow \ (CH_3)_2CO \ + \ CH_3CO \cdot \ + \ V(\underline{IV})$$

On the other hand, acetoin is oxidised by C—H fission in the sequence $CH_3CH(OH)COCH_3 \rightarrow CH_3COCOCH_3 \rightarrow 2 \ CH_3COOH$ without the intervention of acetaldehyde which would be formed as a result of C—C fission. Biacetyl is produced also in the oxidation of acetoin by Fe(III) perchlorate [184] and by alkaline Cu(II) complexes [185].

From a kinetic point of view, the copper(II) oxidation of the ketol α-hydroxyacetophenone has a number of interesting features. In aqueous pyridine, the medium chosen by Wiberg and Nigh [186], the product Cu(I) is strongly complexed and remains in solution. Two Cu(II) ions are consumed for each ketol molecule and phenylglyoxal is the product of reaction.

$$2 \text{ Cu(II)} + C_6H_5COCH_2OH + 2 OH^- = 2 \text{ Cu(I)} + C_6H_5COCHO + 2 H_2O$$

At very low concentrations of Cu(II), the rate is independent of oxidant, being simply

rate = k[ketol]

which suggests that enolisation of the ketol is rate-determining. Wiberg and Nigh verified this by measuring the rate of enolisation directly by nuclear magnetic resonance. At higher concentrations of Cu(II), however, the rate law takes the form

rate = k'[Cu(II)][ketol]

This is interpreted [187] in terms of the formation of a Cu(II)–ketol chelate

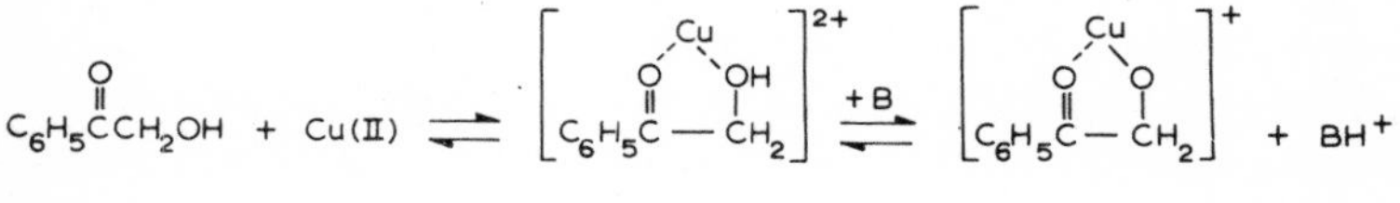

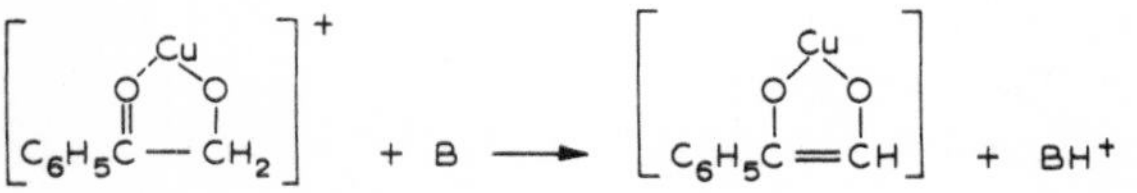

where B represents the pyridine solvent and/or OH⁻ ions. The rate-controlling stage is then considered to be the removal of a proton from the α-methylene group of the complex whereupon the chelate structure is rapidly broken down by

$$\underset{C_6H_5C=CH}{\overset{\overset{\displaystyle Cu}{\underset{O}{|}}\ \overset{\displaystyle}{\underset{O}{|}}}{}} \quad + \quad Cu(II) \quad \longrightarrow \quad \underset{C_6H_5CCHO}{\overset{O}{\overset{\|}{}}} \quad + \quad 2\ Cu(I) \qquad rapid$$

A number of observations lend support to these proposals. Firstly, there is a marked kinetic isotope effect (k_H/k_D = 7.4), the ketol being labelled at the methylene position by acid-catalysed exchange with deuterium oxide. Secondly, the effect of substituents on the rate of oxidation is similar to the effect on the rate of enolisation. Also, copper(II) is known to catalyse the enolisation of other compounds, such as aceto-acetic ester. Furthermore, addition of bipyridine, in sufficient amounts to complex completely with Cu(II), results in a reduced rate approaching that for enolisation.

Early work by Weissberger et al. [188] has shown that the rate of oxidation of benzoin, $C_6H_5CH(OH)COC_6H_5$, to benzil by Cu(II) tartrate in aqueous ethanol is independent of the initial concentration of Cu(II) thus indicating a rate-determining enolisation step. α-Hydroxy aldehydes react in a similar fashion. Marshall and Waters have examined the oxidation of acetoin [185] and benzoin [189] by copper(II) salts. In the case of benzoin, the oxidant was alkaline Cu(II) citrate in a 40% aqueous dioxan medium, citrate being chosen as a means of preventing Cu(I) oxide depositing during the reaction. Although these workers report that oxidation occurs more slowly than does enolisation, this has been criticised by Wiberg and Nigh [186] and it seems likely that in acetoin and benzoin oxidations, enolisation is the rate-determining step. Rate studies of the copper(II) oxidation of various reducing sugars in alkaline media have demonstrated a first-order dependence on the concentration of the sugar and a lack of dependence on Cu(II) concentration, rates of oxidation and enolisation being equal [190]. Oxidation by alkaline ferricyanide is similar [191].

Direct evidence has been cited by Hill and McAuley [192] for an intermediate complex in the reaction between cobalt(III) perchlorate and malic acid, $HOOCCH(OH)CH_2COOH$. Stopped-flow traces reveal an initial increase in absorbance corresponding to the formation of the intermediate, followed by a slower decrease which is associated with an intramolecular electron transfer. The u.v. spectrum of the intermediate has been obtained (Fig. 6) from kinetic measurements made

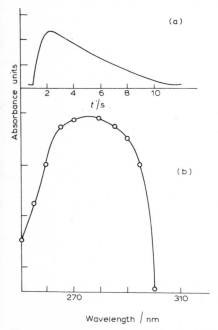

Fig. 6. (a) Stopped-flow trace, 2.4°C, [HA] = 5.02×10^{-2} M, [Co(III)] = 3.9×10^{-4} M, [Co(II)] = 1.14×10^{-3} M, [H$^+$] = 0.14 M. (b) Spectrum of cobalt(III)—malic acid intermediate. (From Hill and McAuley [192], by courtesy of The Chemical Society.)

between 250 and 300 nm. Jørgensen [193] has discussed charge-transfer absorptions in the spectra of metal ion complexes in relation to the oxidising power of the ligand. The peak at 275 nm for the Co(III)—malic acid complex and that at 250 nm for the aquated Co(III) ion are consistent with the view that the more reducing the ligand the greater is the shift of the charge-transfer peak to longer wavelengths. The thiomalic complex peaks at 285 nm, the mercaptan being more strongly reducing than its oxygen analogue [194]. The kinetic data fit the mechanism

$$Co^{3+} + HA \underset{k_{-1}}{\overset{k_1}{\rightleftharpoons}} CoHA^{3+} \qquad K_1$$

$$(H_2O)Co^{3+} \rightleftharpoons CoOH^{2+} + H^+ \qquad K_h$$

$$CoOH^{2+} + HA \underset{k_{-2}}{\overset{k_2}{\rightleftharpoons}} CoOHHA^{2+} \qquad K_2$$

$$(H_2O)CoHA^{3+} \rightleftharpoons CoOHHA^{2+} + H^+ \qquad K_3$$

$$CoHA^{3+} \xrightarrow{k_3} products$$

$$CoOHHA^{2+} \xrightarrow{k_4} products$$

Cobalt(II) was shown not to affect the rates of the complex formation or electron transfer steps. Assuming the electron transfer reactions (k_3 and k_4) to be much slower than the rest, the rate of formation of the intermediate complex is given by

$$\frac{d[CoHA^{3+}]}{dt} = k_1 [Co^{3+}][HA] + k_2 [CoOH^{2+}][HA] - k_{-1}[CoHA^{3+}] - k_{-2}[CoOHHA^{2+}]$$

$$= k_1 [Co^{3+}][HA] + \frac{k_2 K_h}{[H^+]} [Co^{3+}][HA] - k_{-1}[CoHA^{3+}] - \frac{k_{-2} K_3}{[H^+]} [CoHA^{3+}]$$

At equilibrium, $d[CoHA^{3+}]/dt = 0$, and it follows that

$$k_1 = \left[k_{-1} + \frac{k_{-2} K_3}{[H^+]} \right] \frac{\left[CoHA^{3+} \right]_{eq}}{\left[Co^{3+} \right]_{eq} [HA]_{eq}} - \frac{k_2 K_h}{[H^+]}$$

where $[H^+]$ and $[HA]$ are now equilibrium concentrations. The observed first-order rate constant, in the expression rate = $k_{obs} [Co^{3+}]_t$, may be written as

$$k_{obs} = k_{-1} + \frac{k_{-2} K_3}{[H^+]} + \frac{k_{-1} K_1 [HA]}{(1 + K_h/[H^+])} + \frac{k_{-2} K_3 K_1 [HA]}{[H^+] (1 + K_h/[H^+])}$$

which, for constant hydrogen ion concentration, becomes

$$k_{obs} = k_a + k_b [HA]$$

where

$$k_a = k_{-1} + \frac{k_{-2} K_3}{[H^+]}$$

and

$$k_b = \frac{K_1}{(1 + K_h/[H^+])} \left[k_{-1} + \frac{k_{-2} K_3}{[H^+]} \right]$$

Values for k_a and k_b are obtained from the slopes and intercepts of linear plots of k_{obs} versus $[HA]$ at various acidities. Rearrangement of the above equations for k_a and k_b produces

$$k_a / k_b = (K_h / K_1 [H^+]) + 1/K_1$$

and allows K_1 and K_h to be derived from a linear plot of k_a/k_b versus $1/[H^+]$. At 7°C and at an ionic strength of 0.25 M, $K_1 = 34.4 \pm 4$ l mol^{-1} and $K_h = 0.10 \pm 0.05$ mol l^{-1}. Furthermore, a plot of k_a versus $1/[H^+]$ yields k_{-1} as the intercept and $k_{-2}K_3$ as the slope. A value for k_1 (of 5.4 l mol^{-1} s^{-1}) is then obtained knowing k_{-1} and K_1, and a value for k_2 (of 70 l mol^{-1} s^{-1}) from the relationship $k_2 = k_{-2}K_3K_1/K_h$. As regards the electron transfer steps, the rate of oxidation of malic acid may be written

$$\text{rate} = \left[\text{CoHA}^{3+}\right]\left(k_3 + k_4K_3/\left[H^+\right]\right)$$

where $(k_3 + k_4K_3/[H^+])$ is the observed first-order rate constant; k_3 and the composite k_4K_3 are then evaluated from plots of k_{obs} versus $1/[H^+]$.

In sulphuric acid media, hydroxy acids are considered to be oxidised by cerium(IV) through an intervening complex which breaks down to give Ce(III), hydrogen ion, and a radical [195]. The latter then reacts quickly with further Ce(IV). For example, in the case of mandelic acid the sequence is

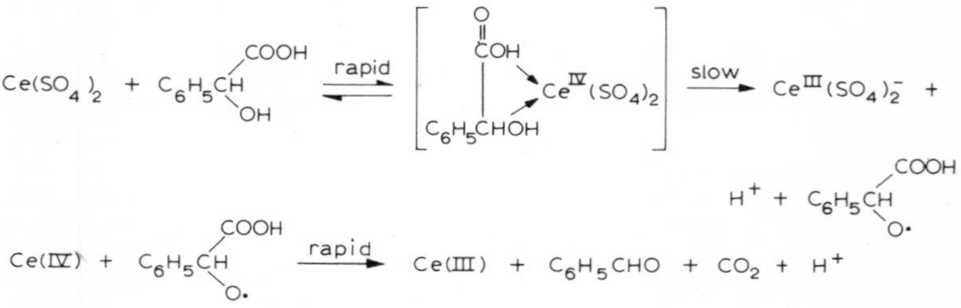

In the above scheme, the reactive species of Ce(IV) is assumed to be Ce(SO$_4$)$_2$ but rate data obtained [196] for the reaction between glycollic acid and Ce(IV) sulphate indicates CeSO$_4^{2+}$.

The oxidation of 2-hydroxy-2-methylpropanoic acid by manganese(III) in aqueous perchloric acid conforms to the stoichiometric equation [197]

$$2\ \text{Mn(III)} + (\text{CH}_3)_2\text{COHCOOH} = 2\ \text{Mn(II)} + (\text{CH}_3)_2\text{CO} + \text{CO}_2 + 2\ \text{H}^+$$

The likely mechanism has a strong resemblance to those for the Ce(IV) + secondary alcohol (p. 50), Ce(IV) + pinacol (p. 55), and Mn(III) + pinacol (p. 55) systems, viz.

$$Mn^{3+} \cdot (CH_3)_2 COHCOOH \; \rightleftharpoons \; Mn^{3+} \cdot (CH_3)_2 COHCOO^- + H^+$$

$$Mn^{3+} \cdot (CH_3)_2 COHCOOH \; \longrightarrow \; Mn(II) + (CH_3)_2 \dot{C}OH + CO_2 + H^+ \quad slow$$

$$Mn^{3+} \cdot (CH_3)_2 COHCOO^- \; \longrightarrow \; Mn(II) + (CH_3)_2 \dot{C}OH + CO_2 \quad slow$$

$$Mn(III) + (CH_3)_2 \dot{C}OH \; \longrightarrow \; Mn(II) + (CH_3)_2 CO + H^+$$

As there, two metal ion—substrate complexes (interrelated by an acid—base equilibrium) are involved.

For the same acid concentration, the rate of oxidation of 2-hydroxy-2-methylpropanoic acid by vanadium(V) is greater in perchloric acid than in sulphuric acid [15] and the reactive species present in the two media are believed to be $V(OH)_3^{2+}$ and $(VO_2, H_2O, H_2SO_4)^+$, respectively, arising from the equilibria

$$V(OH)_4^+ + H^+ \; \rightleftharpoons \; V(OH)_3^{2+} + H_2O \qquad K$$

and

$$V(OH)_4^+ + H_2SO_4 \; \rightleftharpoons \; (VO_2, H_2O, H_2SO_4)^+ + H_2O$$

Kinetically, the reaction is first-order in hydroxy acid and in V(V) concentrations. In the case of perchloric acid solutions, the rate-determining step is likely to be

$$V(OH)_3^{2+} + (CH_3)_2 COHCOOH \; \xrightarrow{k} \; radical \; intermediate + V(IV) + H^+$$

with

$$-\frac{d\left[V(V)\right]}{dt} = \frac{kK\left[V(OH)_4^+\right]\left[hydroxy \; acid\right]\left[H^+\right]}{1 + K\left[H^+\right]}$$

where [V(V)] represents the total V(V) concentration. Since, with hydroxy acid present in excess, the observed rate law is

$$-\frac{d[V(V)]}{dt} = k'\left[V(OH)_4^+\right]$$

it follows that

$$k' = \frac{kK\left[\text{hydroxy acid}\right]\left[H^+\right]}{1 + K\left[H^+\right]}$$

and for constant concentration of organic substrate

$$\frac{1}{k'} = a + \frac{b}{[H^+]}$$

where a and b are constants. The predicted linear relationship between $1/k'$ and $1/[H^+]$ is borne out in practice (Fig. 7). At higher acid concentrations, retardation of the reaction occurs, the rate passing through a maximum at ~ 5 M H_2SO_4 and ~ 7 M $HClO_4$. This effect is attributed to the presence of the equilibria

$$(VO_2, H_2O, H_2SO_4)^+ + H_2SO_4 \rightleftharpoons (VO_2, 2\, H_2SO_4)^+ + H_2O$$

and

$$V(OH)_3^{2+} + H^+ \rightleftharpoons V(OH)_2^{3+} + H_2O$$

which, by shifting to the right at higher acid concentrations, bring about a decrease in the concentration of reactive V(V) species. Lactic, malic, and mandelic acids are oxidised by vanadium(V) according to [198]

$$2\, V(V) + RCHOHCOOH \longrightarrow 2\, V(IV) + RCHO + H_2O + CO_2$$

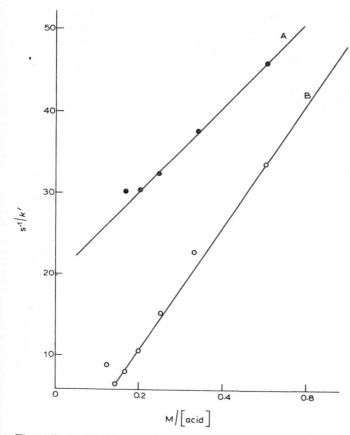

Fig. 7. Plot of reciprocal of observed rate constant versus reciprocal of acid concentration for the reaction between vanadium(V) and 2-hydroxy-2-methylpropanoic acid; A, sulphuric acid and B, perchloric acid. (From Mehrotra [15], by courtesy of The Chemical Society.)

The aldehyde products (acetaldehyde from lactic and malic acids, benzaldehyde from mandelic acid) are then oxidised more slowly.

10. OXIDATION OF PHENOLS

Iron(III) [199], cobalt(III) [200], manganese(III) [201], and cerium(IV) [202] all react quickly with quinol to give p-benzoquinone. The rates of reaction of the first three oxidants are first-order in both metal ion and quinol concentrations, but the rate of oxidation by Ce(IV), although first-order in cation, is independent of the concentration of quinol. Acid dependences vary too: rates of oxidation by Fe(III) and Mn(III) are inversely proportional to [H$^+$] but Co(III) and Ce(IV) oxidations are independent of acidity. Intermediate complexes have been detected (using stopped-flow methods) for Co(III), Mn(III), and Ce(IV) but not for Fe(III). These complexes undergo rapid decomposition to give the semiquinone radical (or its anion) which is then oxidised further by the cation to quinone. The stoichiometry in all cases corresponds to

$$2 M^{n+} + QH_2 = 2 M^{(n-1)+} + Q + 2 H^+$$

In the case of the Co(III) reaction, the mechanism suggested by Wells and Kuritsyn [200] is

$$Co^{3+} + H_2O \rightleftharpoons CoOH^{2+} + H^+ \qquad K_h$$

$$Co^{3+} + QH_2 \rightleftharpoons CoQH_2^{3+} \qquad K$$

$$CoQH_2^{3+} \xrightarrow{k_1} Co(II) + \cdot QH + H^+ \qquad slow$$

$$Co^{3+} + \cdot QH \xrightarrow{k_2} Co(II) + Q + H^+$$

The derived rate law is then

$$-\frac{d[Co(III)]_t}{dt} = \frac{2 k_1 K [Co(III)]_t [QH_2]}{1 + K_h/[H^+] + K [QH_2]}$$

which reduces to that observed if the reasonable assumptions are made that $K[QH_2] \ll 1$ and $K_h/[H^+] \ll 1$ whereupon

$$-\frac{d\left[Co(III)\right]_t}{dt} = 2\,k_1 K\left[Co(III)\right]_t\left[QH_2\right]$$

According to Wells and Kuritsyn [201], oxidation of quinol by Mn(III) takes place by steps analogous to those above with the addition of

$$MnOH^{2+} + QH_2 \rightleftharpoons MnQH^{2+} + H_2O$$
$$MnQH_2^{3+} \rightleftharpoons MnQH^{2+} + H^+$$

and

$$MnQH^{2+} \longrightarrow Mn(II) + \cdot QH$$

but Davies and Kustin [203] find no evidence for complex formation between the reactants. Ce(IV) reacts in a similar fashion [202]. The original suggestion of Baxendale et al. [199] for the Fe(III) + quinol system is

$$Fe^{3+} + H_2O \rightleftharpoons FeOH^{2+} + H^+$$
$$QH_2 \rightleftharpoons QH^- + H^+$$
$$Fe^{3+} + QH^- \rightleftharpoons Fe(II) + \cdot QH$$
$$FeOH^{2+} + QH_2 \rightleftharpoons Fe(II) + \cdot QH + H_2O$$
$$\cdot QH \rightleftharpoons \cdot Q^- + H^+$$
$$Fe(III) + \cdot Q^- \rightleftharpoons Fe(II) + Q$$

The last step is rapid enough to suppress the reaction of semi-quinone with oxygen, i.e.

$$\cdot Q^- + O_2 \longrightarrow Q + \cdot O_2^-$$

or

$$\cdot QH + O_2 \longrightarrow Q + HO_2\cdot$$

and hence the reaction with quinol is not affected by oxygen. Vanadium(V) also oxidises quinol rapidly and the response of the rate to changes in acidity has been explained [204] in terms of the decomposition of the complexes $VO_2QH_2^+$ and $VO(OH)QH_2^{2+}$. A study has been made of the oxidation of quinol and p-toluhydroquinone by neptunium(VI) in aqueous perchlorate media [205]. No evidence is forthcoming that a semiquinone intermediate is of kinetic significance.

Iron(III) oxidises catechol (o-dihydroxybenzene) to o-quinone in aqueous perchloric acid

$$2\ Fe^{3+} + C_6H_4(OH)_2 \rightleftharpoons 2\ Fe^{2+} + C_6H_4O_2 + 2\ H^+$$

via semiquinone radical intermediates although participation of 1:1 chelate complexes cannot be excluded [206a, 206b].

Cecil and Littler [207] have examined the oxidation of phenol and 2,6-dimethylphenol by hexachloroiridate(IV) in acid solution. They conclude that the initial step is an electron transfer from the substrate to the oxidant and that the resulting phenoxyl radical undergoes dimerisation, reacts with further phenol, or is oxidised by a phenoxy cation. Dimerisation explains the formation of the tetramethyldiphenoquinone product.

On mixing perchloric acid solutions of cobalt(III) and sodium salicylate, changes in the absorbance at 400 nm indicate the formation of complexes [208]. Initially, an increase in absorbance occurs due to formation of the monosalicylate complex to be followed by a slower decrease corresponding to the formation of the dicoordinated complex. These complexes are relatively stable but electron transfer does take place slowly and Co(III) is reduced to Co(II). It is likely that the major organic product is muconic acid.

11. OXIDATION OF ETHERS

Kinetic and product studies of the oxidation of three benzyl ethers and diphenyl ether by cobalt(III) have been reported by Cooper and Waters [209], the limited solubility of the compounds necessitating a mixed solvent of methyl cyanide and water. Since the reaction rates are low, complications arise from the thermal decomposition of Co(III) itself. A direct attack of the oxidant on the ether is indicated by the first-order dependence of the rate on the concentration of each reactant. In the case of dibenzyl ether (in 60% CH_3CN solution), benzaldehyde is the major product along with benzoic acid and lesser amounts of benzyl alcohol. The following sequence of reactions accounts for the observed products.

$$Co(III) + C_6H_5CH_2OCH_2C_6H_5 \longrightarrow Co(II) + C_6H_5\dot{C}HOCH_2C_6H_5 + H^+$$

$$C_6H_5\dot{C}HOCH_2C_6H_5 \longrightarrow C_6H_5CHO + C_6H_5CH_2^{\cdot}$$

$$Co(III) + C_6H_5CH_2^{\cdot} + H_2O \longrightarrow Co(II) + C_6H_5CH_2OH + H^+$$

$$2\ Co(III) + C_6H_5CH_2OH \longrightarrow 2\ Co(II) + C_6H_5CHO + 2\ H^+$$

$$2\ Co(III) + C_6H_5CHO + H_2O \longrightarrow 2\ Co(II) + C_6H_5COOH + 2\ H^+$$

The presence of trace quantities of bibenzyl in the products is supporting evidence for the participation of the $C_6H_5CH_2^{\cdot}$ radical. Furthermore, traces of benzyl benzoate are detected and may arise from the competing reaction of the $C_6H_5\overset{\cdot}{C}HOCH_2C_6H_5$ radical with oxidant, viz.

$$Co(III) + C_6H_5\overset{\cdot}{C}HOCH_2C_6H_5 + H_2O \longrightarrow Co(II) + C_6H_5CH(OH)OCH_2C_6H_5 + H$$

$$2\,Co(III) + C_6H_5CH(OH)OCH_2C_6H_5 \longrightarrow 2\,Co(II) + C_6H_5COOCH_2C_6H_5 + 2\,H^+$$

An additional reaction of this radical is possibly

$$Co(III) + C_6H_5\overset{\cdot}{C}HOCH_2C_6H_5 + H_2O \longrightarrow Co(II) + C_6H_5CHO + C_6H_5CH_2OH +$$

The ethers investigated follow the reactivity sequence $C_6H_5OC_6H_5 > C_6H_5CH_2OC_6H_{11} > C_6H_5CH_2OCH_2C_6H_5 > C_6H_5CH_2OCH_3$. As is usual for Co(III) oxidations, rates vary inversely with acidity. The increase in rate, as the proportion of methyl cyanide in the solvent is increased, is attributed to the formation of more reactive $Co(III)–CH_3CN$ complexes (as the CH_3CN concentration is increased, the colour of a Co(III) solution in 1 M perchloric acid changes from blue-green to blue to violet to cherry-red).

Oxidation of p-methoxytoluene (PMT) by manganese(III) acetate in acetic acid has been examined in detail by Andrulis et al. [19], the overall reaction corresponding to

$$2\,Mn(OAc)_3 + MeOC_6H_4CH_3 = 2\,Mn(OAc)_2 + HOAc + MeOC_6H_4CH_2OAc$$

In the presence of oxygen, anisyl acetate is the main product (as it is when Co(III) and Pb(IV) are the oxidants) although some anisaldehyde is formed. However, chromium trioxide and permanganate produce considerable amounts of anisic acid. Under anaerobic conditions at 70—100°C, the rate law is

$$-\frac{d\left[Mn(III)\right]}{dt} = \frac{k\left[Mn(III)\right]\left[PMT\right]}{\left[Mn(II)\right]}$$

A scheme based upon the disproportionation of Mn(III)

$$2\ Mn(III) \rightleftharpoons Mn(II) + Mn(IV)$$

$$Mn(IV) + PMT \longrightarrow products$$

is inconsistent with the kinetic data since a first-order dependence on Mn(III) concentration would not result. Instead, the reaction is viewed as proceeding via an organic intermediate X, i.e.

$$Mn(III) + PMT \rightleftharpoons Mn(II) + X$$

$$X \longrightarrow products$$

As the reaction shows a large deuterium isotope effect, the implication is that X is a radical ion derived from the substrate by electron transfer and that the slow stage is the loss of a proton from X to yield an anisyl radical. The latter is oxidised by Mn(III) to the corresponding carbonium ion which then reacts with solvent to give anisyl acetate.

$$Mn(III) + MeOC_6H_4CH_3 \rightleftharpoons Mn(II) + MeOC_6H_4CH_3^+$$

$$MeOC_6H_4CH_3^+ \longrightarrow MeOC_6H_4\dot{C}H_2 + H^+ \qquad slow$$

$$Mn(III) + MeOC_6H_4\dot{C}H_2 \longrightarrow Mn(II) + MeOC_6H_4CH_2^+ \qquad rapid$$

$$MeOC_6H_4CH_2^+ + HOAc \longrightarrow MeOC_6H_4CH_2OAc + H^+ \qquad rapid$$

To rationalise the lack of effect of sodium acetate on the reaction rate, the initial stages can be detailed as

$$Mn(OAc)_3 + MeOC_6H_4CH_3 \rightleftharpoons Mn(OAc)_2 + (MeOC_6H_4CH_3^+, OAc^-)$$
$$\text{ion-pair}$$

$$(MeOC_6H_4CH_3^+, OAc^-) \longrightarrow MeOC_6H_4\overset{\cdot}{C}H_2 + HOAc$$

since Mn(III) acetate is undissociated in acetic acid and ion-pairs are favoured by the low dielectric constant of this medium. A similar scheme has been applied to the oxidation of 1- and 2-methoxynaphthalene [210].

12. OXIDATION OF SULPHUR COMPOUNDS

Thiourea and substituted thioureas are considered to react with cobalt(III) perchlorate via intermediate complexes, the formation of which is rate-determining [211a, 211b]. Although direct spectrophotometric evidence for the intermediates is not forthcoming, comparison with the rates of other Co(III) oxidations, e.g. of malic and thiomalic acids, suggests the reaction to be essentially inner-sphere in character. The rate is first-order in the concentration of each reactant and the kinetics are compatible with the mechanism

$$Co^{3+} + RSH^+ \longrightarrow Co^{2+} + RS^+ + H^+ \qquad \text{slow}$$
$$CoOH^{2+} + RSH^+ \longrightarrow Co^{2+} + RS^+ + H_2O \qquad \text{slow}$$
$$2\ RS^+ \longrightarrow RSSR^{2+} \qquad \text{rapid}$$

where thiourea is present largely as the protonated species and the product is a disulphide. The oxidation of thiourea and its N-substituted derivatives by cerium(IV) sulphate follows the reactivity pattern [212] dimethyl- > diethyl- > thiourea > ethylenethiourea. As expected, this sequence correlates with the order of inductive effects; the greater the tendency to increase the electron density at the point of transfer the greater is the rate of reduction of Ce(IV). The small variation of rate in the reaction of $MnOH^{2+}$ with mono-protonated thioureas has led Davies [213] to suggest that the reactions are essentially substitution-controlled processes, the overall rate being controlled by the rate of conversion of the ion pair (Mn(III),B) into the inner-sphere complex (Mn(III)B)

$$Mn(III) + B \rightleftharpoons Mn(III), B \qquad K_0$$

$$Mn(III), B \xrightleftharpoons{k_0} Mn(III)B + H_2O$$
$$\downarrow$$
$$product$$

with $k_{obs} = k_0 K_0$.

Alkaline ferricyanide oxidises thiourea and thioacetamide according to the stoichiometric equations [214]

$$8 \; Fe(CN)_6^{3-} + NH_2 CSNH_2 + 10 \; OH^- = 8 \; Fe(CN)_6^{4-} + NH_2CONH_2 + SO_4^{2-} + 5 \; H_2O$$

$$8 \; Fe(CN)_6^{3-} + CH_3CSNH_2 + 11 \; OH^- = 8 \; Fe(CN)_6^{4-} + CH_3COO^- + SO_4^{2-} + NH_3 + 5 \; H_2O$$

At pH 10 (in carbonate—bicarbonate buffers), oxidation of thiourea is first-order in thiourea, $Fe(CN)_6^{3-}$ and OH^- concentrations, whereas oxidation of thioacetamide is first-order in thioacetamide and OH^- ion but independent of $Fe(CN)_6^{3-}$. · The observed kinetics are consistent with the general scheme

$$RCSNH_2 + OH^- \xrightleftharpoons[k_{-1}]{k_1} RCSNH^- + H_2O$$

$$Fe(CN)_6^{3-} + RCSNH^- \xrightarrow{k_2} complex$$

$$Fe(CN)_6^{3-} + complex \longrightarrow products \qquad rapid$$

and, applying the steady-state approximation to the concentration of the intermediate anion, the derived rate law becomes

$$-\frac{d\left[Fe(CN)_6^{3-}\right]}{dt} = \frac{2\,k_1\,k_2\left[RCSNH_2\right]\left[OH^-\right]\left[Fe(CN)_6^{3-}\right]}{k_{-1} + k_2\left[Fe(CN)_6^{3-}\right]}$$

If $k_{-1} \gg k_2$, the rate law reduces to

$$-\frac{d\left[Fe(CN)_6^{3-}\right]}{dt} = \frac{2\,k_1\,k_2}{k_{-1}}\left[RCSNH_2\right]\left[OH^-\right]\left[Fe(CN)_6^{3-}\right]$$

and the second stage is rate-determining as is the case for the oxidation of thiourea. If, on the other hand, step k_1 is rate-determining and $k_{-1} \ll k_2\,[Fe(CN)_6^{3-}]$, then

$$-\frac{d\left[Fe(CN)_6^{3-}\right]}{dt} = 2\,k_1\left[RCSNH_2\right]\left[OH^-\right]$$

as is found for thioacetamide oxidation. In the latter reaction, the rate becomes dependent on $Fe(CN)_6^{3-}$ as the oxidant concentration is reduced.

A stopped-flow study of the oxidation of some sulphur analogues of carboxylic acids by cerium(IV) has been made [215]. Rates of oxidation of thiolactic, thiomalic, and thioglycollic acids in sulphuric acid are first-order in the concentration of each reactant, kinetic measurements being made on the rate of disappearance of Ce(IV) at 300—320 nm. Under conditions of excess substrate, the products are disulphides, each mole of Ce(IV) consuming approximately one mole of mercaptocarboxylic acid. (If Ce(IV) is in excess, then the disulphides suffer further oxidation to sulphoxides and sulphones.) No significant complex formation takes place before electron transfer and the reaction scheme is essentially

$$Ce(SO_4)_n^{(4-2n)+} + RSH \longrightarrow Ce(III) + RS\cdot + H^+$$

$$2\ RS\cdot \longrightarrow RSSR$$

It has been reported that iodine oxidises thiols beyond the disulphide stage and this has also been noted for the oxidation of mercaptocarboxylic acids by Ce(IV) and Np(VI) under anaerobic conditions in sulphuric acid and perchloric acid solutions, respectively [216].

Vanadium(V) oxidises 2-mercaptosuccinic acid via a coloured intermediate complex, whose formation and decay can be followed by stopped-flow techniques [217]. In Fig. 8 are some typical oscillograph traces; the initial fall in the curves arises from the formation of the complex (VT). The minimum absorbance corresponds to the equilibrium concen-

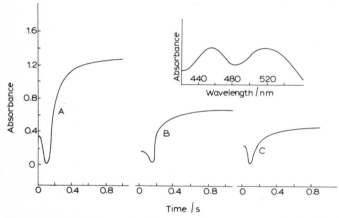

Fig. 8. Oscillograph traces showing the formation and decomposition of the coloured intermediate (absorption spectrum in inset); 10^{-3} M NH_4VO_3 reacting with A, 5×10^{-2} M; B, 2.5×10^{-2} M; and C, 10^{-2} M mercaptosuccinic acid. (From Pickering and McAuley [217], by courtesy of The Chemical Society.)

tration of the complex ($[VT]_{eq}$, time t_{eq}) and, with the thiol (T) in excess, the rate of formation of VT is first-order such that plots of $\log([VT]_{eq} - [VT])$ versus $(t_{eq} - t)$ are linear with slopes equal to k'. For the equilibrium

$$V + T \; \underset{k_2}{\overset{k_1}{\rightleftharpoons}} \; VT$$

where T is in large excess over V

$$\frac{d[VT]}{dt} = \left([VT]_{eq} - [VT]\right)(k_2 + k_1[T])$$

and therefore

$$k' = k_2 + k_1[T]$$

That is, k' varies linearly with thiol concentration, thus allowing k_1, k_2, and the formation constant to be calculated. Pickering and McAuley [217] advocate the mechanism

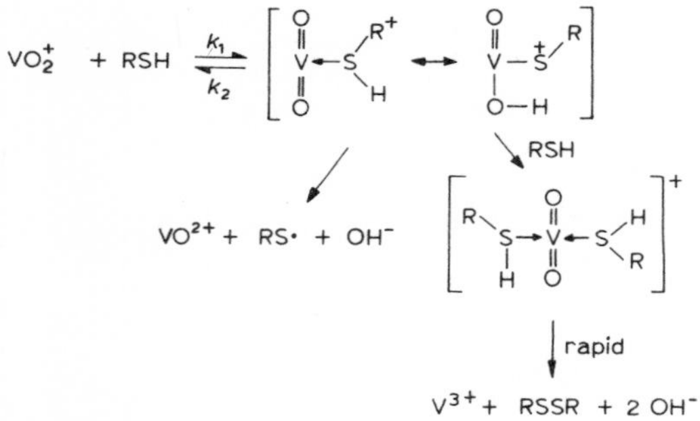

along with

$$2\,RS\cdot \longrightarrow RSSR$$

and

$$VO_2^+ + V^{3+} \longrightarrow 2 \, VO^{2+}$$

Thiomalic acid (2-mercaptosuccinic acid) is oxidised by Fe(III) through a blue transient complex which has an absorption maximum at 610 nm [218]. By way of contrast, the complex formed between malic acid and Fe(III) is yellow in colour. The ferricyanide oxidations of 3-mercaptopropionic acid [219] and thioglycollic acid [220] are similar in that they are both retarded by ferrocyanide, indicating the initial radical-producing steps to be reversible.

REFERENCES

1   G. Hargreaves and L.H. Sutcliffe, Trans. Faraday Soc., 51 (1955) 786.
2   M. Anbar and I. Pecht, J. Amer. Chem. Soc., 89 (1967) 2553.
3   L.H. Sutcliffe and J.R. Weber, J. Inorg. Nucl. Chem., 12 (1960) 281.
4a B. Warnqvist, Inorg. Chem., 9 (1970) 682.
4b G. Davies and B. Warnqvist, J. Chem. Soc. Dalton, (1973) 900.
5   H.L. Friedman, J.P. Hunt, R.A. Plane and H. Taube, J. Amer. Chem. Soc., 73 (1951) 4028.
6   J.H. Baxendale and C.F. Wells, Trans. Faraday Soc., 53 (1957) 800.
7   G. Davies and B. Warnqvist, Coord. Chem. Rev., 5 (1970) 349.
8   D.H. Huchital, N. Sutin and B. Warnqvist, Inorg. Chem., 6 (1967) 838.
9   M.G. Adamson, F.S. Dainton and P. Glentworth, Trans. Faraday Soc., 61 (1965) 689.
10 T.J. Hardwick and E. Robertson, Can. J. Chem., 29 (1951) 828.
11 M.S. Sherrill, C.G. King and R.C. Spooner, J. Amer. Chem. Soc., 65 (1943) 170.
12a L.J. Heidt and M.E. Smith, J. Amer. Chem. Soc., 70 (1948) 2476.
12b M. Ardon and G. Stein, J. Chem. Soc. London, (1956) 104.
13 F.B. Baker, T.W. Newton and M. Kahn, J. Phys. Chem., 64 (1960) 109.

104

14 J.S. Littler and W.A. Waters, J. Chem. Soc. London, (1959) 4046.
15 R.N. Mehrotra, J. Chem. Soc. B, (1968) 642.
16 H. Diebler and N. Sutin, J. Phys. Chem., 68 (1964) 174.
17 R.G. Selim and J.J. Lingane, Anal. Chim. Acta, 21 (1959) 536.
18 D.R. Rosseinsky, J. Chem. Soc. London, (1963) 1181.
19 P.J. Andrulis, M.J.S. Dewar, R. Dietz and R.L. Hunt, J. Amer. Chem. Soc., 88 (1966) 5473.
20 C.F. Wells and G. Davies, J. Chem. Soc. A, (1967) 1858.
21 G. Davies, Coord. Chem. Rev., 4 (1969) 199.
22 L. Ciavatta and M. Grimaldi, J. Inorg. Nucl. Chem., 31 (1969) 3071.
23 K.J. Vetter and G. Manecke, Z. Physik. Chem., 195 (1950) 270.
24 J.I. Watters and I.M. Kolthoff, J. Amer. Chem. Soc., 70 (1948) 2455.
25 M.A. Suwyn and R.E. Hamm, Inorg. Chem., 6 (1967) 142.
26 J.B. Kirwin, F.D. Peat, P.J. Proll and L.H. Sutcliffe, J. Phys. Chem., 67 (1963) 1617.
27 J.B. Kirwin, P.J. Proll and L.H. Sutcliffe, Trans. Faraday Soc., 60 (1964) 119.
28 R.N. Hammer and J. Kleinberg, Inorg. Syn., 4 (1953) 12.
29 G. Davies, Inorg. Chem., 10 (1971) 1155.
30 L.E. Bennett and J.C. Sheppard, J. Phys. Chem., 66 (1962) 1275.
31 M.R. Hyde, R. Davies and A.G. Sykes, J. Chem. Soc. Dalton, (1972) 1838.
32a T.J. Conocchioli, G.H. Nancollas and N. Sutin, J. Amer. Chem. Soc., 86 (1964) 1453.
32b D.R. Rosseinsky and W.C.E. Higginson, J. Chem. Soc. London, (1960) 31.
33 K.G. Ashurst and W.C.E. Higginson, J. Chem. Soc. London, (1956) 343.
34a L.H. Sutcliffe and J.R. Weber, Trans. Faraday Soc., 52 (1956) 1225.
34b L.H. Sutcliffe and J.R. Weber, Trans. Faraday Soc., 55 (1959) 1892.
34c L.H. Sutcliffe and J.R. Weber, Trans. Faraday Soc., 57 (1961) 91.
35 J.C. Sullivan and R.C. Thompson, Inorg. Chem., 6 (1967) 1795.
36 J.Y.P. Tong and E.L. King, J. Amer. Chem. Soc., 82 (1960) 3805.
37 G.A. Rechnitz, G.N. Rao and G.P. Rao, Anal. Chem., 38 (1966) 1900.
38 G. Dulz and N. Sutin, Inorg. Chem., 2 (1963) 917.
39 R.A. Marcus, Discuss. Faraday Soc., 29 (1960) 21.
40 P.A. Rodriguez and H.L. Pardue, Anal. Chem., 41 (1969) 1369.
41 P.A. Rodriguez and H.L. Pardue, Anal. Chem., 41 (1969) 1376.
42 S.K. Mishra and Y.K. Gupta, J. Chem. Soc. A, (1970) 260.

43a   N.A. Daugherty and T.W. Newton, J. Phys. Chem., 67 (1963) 1090.
43b   D.R. Rosseinsky and M.J. Nicol, Electrochim. Acta, 11 (1966) 1069.
44   J.P. Birk and T.P. Logan, Inorg. Chem., 12 (1973) 580.
45a   D.J. Drye, W.C.E. Higginson and P. Knowles, J. Chem. Soc. London, (1962) 1137.
45b   B. Schiefelbein and N.A. Daugherty, Inorg. Chem., 9 (1970) 1716.
46   J.P. Birk, Inorg. Chem., 9 (1970) 125.
47   J.H. Espenson, Inorg. Chem., 7 (1968) 631.
48   J.H. Espenson and L.A. Krug, Inorg. Chem., 8 (1969) 2633.
49   N.A. Daugherty and T.W. Newton, J. Phys. Chem., 68 (1964) 612.
50   C.R. Guiliano and H.M. McConnell, J. Inorg. Nucl. Chem., 9 (1959) 171.
51   D.R. Rosseinsky and M.J. Nicol, J. Chem., Soc. A, (1968) 1022.
52   D.R. Rosseinsky and R.J. Hill, J. Chem. Soc. Dalton, (1972) 715.
53   W.C.E. Higginson and A.G. Sykes, J. Chem. Soc. London, (1962) 2841.
54   G. Dulz and N. Sutin, J. Amer. Chem. Soc., 86 (1964) 829.
55a   D.W. Carlyle and J.H. Espenson, Inorg. Chem., 8 (1969) 575.
55b   D.W. Carlyle and J.H. Espenson, J. Amer. Chem. Soc., 91 (1969) 599.
56   A. Haim and W.K. Wilmarth, J. Amer. Chem. Soc., 83 (1961) 509.
57   R.N.F. Thorneley and A.G. Sykes, J. Chem. Soc. A, (1970) 862.
58   B.R. Baker, M. Orhanovic and N. Sutin, J. Amer. Chem. Soc., 89 (1967) 722.
59   F.R. Duke and R.C. Pinkerton, J. Amer. Chem. Soc., 73 (1951) 3045.
60   D.W. Carlyle and J.H. Espenson, J. Amer. Chem. Soc., 90 (1968) 2272.
61   R.H. Betts, Can. J. Chem., 33 (1955) 1780.
62   T.W. Newton, G.E. McCrary and W.G. Clark, J. Phys. Chem., 72 (1968) 4333.
63   J.R. Huizenga and L.B. Magnusson, J. Amer. Chem. Soc., 73 (1951) 3202.
64   R.S. Taylor and A.G. Sykes, J. Chem. Soc. A, (1971) 1628.
65   O.J. Parker and J.H. Espenson, Inorg. Chem., 8 (1969) 1523.
66   K. Shaw and J.H. Espenson, Inorg. Chem., 7 (1968) 1619.
67   O.J. Parker and J.H. Espenson, Inorg. Chem., 8 (1969) 185.
68   T.L. Nunes, Inorg. Chem., 9 (1970) 1325.
69   D. Meyerstein, Inorg. Chem., 10 (1971) 638.
70   M. Anbar, Advan. Chem. Ser., 49 (1965) 126.

71   J.B. Kirwin, F.D. Peat, P.J. Proll and L.H. Sutcliffe, J. Phys. Chem., 67 (1963) 2288.
72   W.C.E. Higginson, D.R. Rosseinsky, J.B. Stead and A.G. Sykes, Discuss. Faraday Soc., 29 (1960) 49.
73   F.R. Duke and C.E. Borchers, J. Amer. Chem. Soc., 75 (1953) 5186.
74   A. McAuley, M.N. Malik and J. Hill, J. Chem. Soc. A, (1970) 2461.
75   E.L. King and M.L. Pandow, J. Amer. Chem. Soc., 75 (1953) 3063.
76a   M.N. Malik, J. Hill and A. McAuley, J. Chem. Soc. A, (1970) 643.
76b   G. Davies and K.O. Watkins, J. Phys. Chem., 74 (1970) 3388.
77   C.F. Wells and D. Mays, J. Chem. Soc. A, (1968) 577.
78   K. Julian and W.A. Waters, J. Chem. Soc. London, (1962) 818.
79   G. St. Nikolov and D. Mihailova, J. Inorg. Nucl. Chem., 31 (1969) 2499.
80a   G.S. Laurence and K.J. Ellis, J. Chem. Soc. Dalton, (1972) 2229.
80b   A.J. Fudge and K.W. Sykes, J. Chem. Soc. London, (1952) 119.
81   F. Secco, C. Celsi and C.Grati, J. Chem. Soc. Dalton, (1972) 1675.
82   Y.A. Majid and K.E. Howlett, J. Chem. Soc. A, (1968) 679.
83   C.F. Wells and D. Mays, J. Chem. Soc. A, (1968) 2740.
84   R.C. Thompson, J. Phys. Chem., 72 (1968) 2642.
85   C.F. Wells and D. Mays, J. Chem. Soc. A, (1969) 2175.
86a   R.K. Murmann, J.C. Sullivan and R.C. Thompson, Inorg. Chem., 7 (1968) 1876.
86b   R.C. Thompson and J.C. Sullivan, Inorg. Chem., 9 (1970) 1590.
87a   C.F. Wells and D. Mays, J. Chem. Soc. A, (1968) 1622.
87b   G. Davies, L.J. Kirschenbaum and K. Kustin, Inorg. Chem., 8 (1969) 663.
88   C.F. Wells and M. Husain, J. Chem. Soc. A, (1969) 2981.
89   J.M. Lancaster and R.S. Murray, J. Chem. Soc. A, (1971) 2755.
90   R.H. Betts and F.S. Dainton, J. Amer. Chem. Soc., 75 (1953) 5721.
91   F.M. Page, Trans. Faraday Soc., 56 (1960) 398.
92   W.C.E. Higginson and J.W. Marshall, J. Chem. Soc. London, (1957) 447.
93   J.J. Byerley, S.A. Fouda and G.L. Rempel, J. Chem. Soc. Dalton, (1973) 889.
94   G. Veith, E. Guthals and A. Viste, Inorg. Chem., 6 (1967) 667.
95   A. Viste, D.A. Holm, P.L. Wang and G.D. Veith, Inorg. Chem., 10 (1971) 631.
96   J.N. Cooper, H.L. Hoyt, C.W. Buffington and C.A. Holmes, J. Phys. Chem., 75 (1971) 891.
97a   E. Peters and J. Halpern, J. Phys. Chem., 59 (1955) 793.

97b   J. Halpern, E.R. MacGregor and E. Peters, J. Phys. Chem., 60
       (1956) 1455.

98   A.H. Webster and J. Halpern, J. Phys. Chem., 61 (1957) 1239.

99   C.F. Wells and M. Husain, Trans. Faraday Soc., 67 (1971) 760.

100a   C.F. Wells and M. Husain, J. Chem. Soc. A, (1970) 1013.

100b   H.A. Mahlman, R.W. Matthews and T.J. Sworski, J. Phys. Chem.,
       75 (1971) 250.

101a   C.F. Wells and D. Mays, J. Chem. Soc. A, (1968) 665.

101b   G. Davies, L.K. Kirschenbaum and K. Kustin, Inorg. Chem., 7
       (1968) 146.

102   M.G. Evans, P. George and N. Uri, Trans. Faraday Soc., 45 (1949)
       230.

103   C.F. Wells and D. Mays, Inorg. Nucl. Chem. Lett., 5 (1969) 9.

104   A.E. Cahill and H. Taube, J. Amer. Chem. Soc., 74 (1952) 2312.

105   E. Saito and B.H.J. Bielski, J. Phys. Chem., 66 (1962) 2266.

106   A. Samuni and G. Czapski, J. Chem. Soc. Dalton, (1973) 487.

107   J.I. Morrow and L. Silver, Inorg. Chem., 11 (1972) 231.

108   G. Davies and K. Kustin, Inorg. Chem., 8 (1969) 484.

109   D.S. Honig, K. Kustin and J.F. Martin, Inorg. Chem., 11 (1972)
       1895.

110   V.K. Jindal, M.C. Agrawal and S.P. Mushran, J. Chem. Soc. A,
       (1970) 2060.

111   G. Davies and K. Kustin, J. Phys. Chem., 73 (1969) 2248.

112   J.I. Morrow and G.W. Sheeres, Inorg. Chem., 11 (1972) 2606.

113   J. Hanotier, P. Camerman, M. Hanotier-Bridoux and P. De
       Radzitzky, J. Chem. Soc. Perkin II, (1972) 2247.

114   C.F. Wells, Trans. Faraday Soc., 63 (1967) 156.

115   T.A. Cooper and W.A. Waters, J. Chem. Soc. B, (1967) 687.

116a   E.I. Heiba, R.M. Dessau and W.J. Koehl, J. Amer. Chem. Soc.,
       91 (1969) 6830.

116b   K. Sakota, Y. Kamiya and N. Ohta, Can. J. Chem., 47 (1969) 387.

117   E.I. Heiba, R.M. Dessau and W.J. Koehl, J. Amer. Chem. Soc.,
       91 (1969) 138.

118   E.I. Heiba, R.M. Dessau and W.J. Koehl, J. Amer. Chem. Soc.,
       90 (1968) 1082.

119   J. Hanotier and M. Hanotier-Bridoux, J. Chem. Soc. Perkin II,
       (1973) 1035.

120   J. Hanotier, M. Hanotier-Bridoux and P. De Radzitzky, J. Chem.
       Soc. Perkin II, (1973) 381.

121   P.S.R. Murti and S.C. Pati, Chem. Ind. London, (1966) 1722.

122   P.S.R. Murti and S.C. Pati, Chem. Ind. London, (1967) 702.
123a   C.E.H. Bawn and J.A. Sharp, J. Chem. Soc. London, (1957) 1854.
123b   C.E.H. Bawn and J.A. Sharp, J. Chem. Soc. London, (1957) 1866.
124   C.E.H. Bawn and A.G. White, J. Chem. Soc. London, (1951) 343.
125   M. Ardon, J. Chem. Soc. London, (1957) 1811.
126   S.S. Muhammad and K.U. Rao, Bull. Chem. Soc. Japan, 36 (1963) 943.
127   C.F. Wells and C. Barnes, J. Chem. Soc. A, (1968) 1626.
128   C.F. Wells and C. Barnes, J. Chem. Soc. A, (1971) 430.
129   J.R. Jones and W.A. Waters, J. Chem. Soc. London, (1960) 2772.
130   C.F. Wells and M. Husain, Trans. Faraday Soc., 66 (1970) 679.
131a   C.F. Wells and M. Husain, Trans. Faraday Soc., 66 (1970) 2855.
131b   H.L. Hintz and D.C. Johnson, J. Org. Chem., 32 (1967) 556.
132a   C.F. Wells and G. Davies, Trans. Faraday Soc., 63 (1967) 2737.
132b   C.F. Wells, C. Barnes and G. Davies, Trans. Faraday Soc., 64 (1968) 3069.
133   K. Meyer and J. Roček, J. Amer. Chem. Soc., 95 (1972) 1209.
134a   D.G. Hoare and W.A. Waters, J. Chem. Soc. London, (1962) 965.
134b   D.G. Hoare and W.A. Waters, J. Chem. Soc. London, (1964) 2552.
135   J.R. Jones and W.A. Waters, J. Chem. Soc. London, (1962) 2068.
136   H. Land and W.A. Waters, J. Chem. Soc. London, (1958) 2129.
137   G.G. Guilbault and W.H. McCurdy, J. Phys. Chem., 67 (1963) 283.
138   F.R. Duke and R.F. Bremer, J. Amer. Chem. Soc., 73 (1951) 5179.
139   J.S. Littler and W.A. Waters, J. Chem. Soc. London, (1960) 2767.
140   W.S. Trahanovsky, L.H. Young and M.H. Bierman, J. Org. Chem., 34 (1969) 869.
141   C.F. Wells and C. Barnes, J. Chem. Soc. A, (1971) 1405.
142   C.F. Wells and M. Husain, Trans. Faraday Soc., 67 (1971) 1086.
143   J.S. Littler and W.A. Waters, J. Chem. Soc. London, (1959) 1299.
144   G. Hargreaves and L.H. Sutcliffe, Trans. Faraday Soc., 51 (1955) 1105.
145   T.J. Kemp and W.A. Waters, J. Chem. Soc. London, (1964) 339.
146   T.J. Kemp and W.A. Waters, Proc. Roy. Soc. London Ser. A, 274 (1963) 480.
147   V.N. Singh, M.C. Gangwar, B.B.L. Saxena and M.P. Singh, Can. J. Chem., 47 (1969) 1051.
148   U. Shanker and M.P. Singh, Indian J. Chem., 6 (1968) 702.
149   J.R. Jones and W.A. Waters, J. Chem. Soc. London, (1963) 352.
150   A. Lorenzini and C. Walling, J. Org. Chem., 32 (1967) 4008.
151   C.E. Castro, E.J. Gaughan and D.C. Owsley, J. Org. Chem., 30 (1965) 587.

152   W.G. Nigh, in W.S. Trahanovsky (Ed.), Oxidation in Organic
       Chemistry, Part B, Academic Press, New York, 1973, p. 78.
153   T.A. Cooper and W.A. Waters, J. Chem. Soc. London, (1964) 1538.
154   C.E.H. Bawn and J.E. Jolley, Proc. Roy. Soc. London Ser. A, 237
       (1956) 313.
155   K.B. Wiberg and P.C. Ford, J. Amer. Chem. Soc., 91 (1969) 124.
156   K.B. Wiberg and P.C. Ford, Inorg. Chem., 7 (1968) 369.
157a   J. Shorter and C.N. Hinshelwood, J. Chem. Soc. London, (1950)
       3276.
157b   J. Shorter and C.N. Hinshelwood, J. Chem. Soc. London, (1950) 3425.
158   J. Shorter, J. Chem. Soc. London, (1962) 1868.
159   J.S. Littler, J. Chem. Soc. London, (1962) 832.
160   J.S. Littler, J. Chem. Soc. London, (1962) 827.
161   D.G. Hoare and W.A. Waters, J. Chem. Soc. London, (1962) 971.
162   J.S. Littler and I.G. Sayce, J. Chem. Soc. London, (1964) 2545.
163   R. Cecil, J.S. Littler and G. Easton, J. Chem. Soc. B, (1970) 626.
164   J.K. Kochi, J. Amer. Chem. Soc., 77 (1955) 5274.
165   W.G. Nigh, in W.S. Trahanovsky (Ed.), Oxidation in Organic Chem-
       istry, Part B, Academic Press, New York, 1973, p. 76.
166   C.F. Wells and D. Whatley, J. Chem. Soc. Faraday I, 68 (1972) 434.
167   C.F. Wells and M. Husain, J. Chem. Soc. A, (1971) 380.
168   A.A. Clifford and W.A. Waters, J. Chem. Soc. London, (1965)
       2796.
169   T.A. Cooper, A.A. Clifford, D.J. Mills and W.A. Waters, J. Chem.
       Soc. C, (1966) 793.
170   P.R. Sharan, P. Smith and W.A. Waters, J. Chem. Soc. B, (1968)
       1322.
171   I.M. Mathai and R. Vasudevan, J. Chem. Soc. B, (1970) 1361.
172   S.S. Lande and J.K. Kochi, J. Amer. Chem. Soc., 90 (1968) 5196.
173   R.E. van der Ploeg, R.W. de Korte and E.C. Kooyman, J. Catal.,
       10 (1968) 52.
174   D. Greatorex and T.J. Kemp, Chem. Commun., (1969) 383.
175   C.F. Wells and C. Barnes, Trans. Faraday Soc., 66 (1970) 1154.
176a   G. Davies and K.O. Watkins, Inorg. Chem., 9 (1970) 2735.
176b   Y.A. El-Tantawy and G.A. Rechnitz, Anal. Chem., 36 (1964) 1774.
177   G.A. Rechnitz and Y.A. El-Tantawy, Anal. Chem., 36 (1964) 2361.
178   J.R. Jones and W.A. Waters, J. Chem. Soc. London, (1961) 4757.
179   T.J. Kemp and W.A. Waters, J. Chem. Soc. London, (1964) 3101.
180   N.K. Shastri and E.S. Amis, Inorg. Chem., 8 (1969) 2487.
181   H. Degn, Nature London, 213 (1967) 589.
182   G.J. Kasperek and T.C. Bruice, Inorg. Chem., 10 (1971) 382.
183   J.R. Jones and W.A. Waters, J. Chem. Soc. London, (1962) 1629.

184   J.K. Thomas, G. Trudel and S. Bywater, J. Phys. Chem., 64 (1960) 51.
185   B.A. Marshall and W.A. Waters, J. Chem. Soc. London, (1960) 2392.
186   K.B. Wiberg and W.G. Nigh, J. Amer. Chem. Soc., 87 (1965) 3849.
187   W.G. Nigh, in W.S. Trahanovsky (Ed.), Oxidation in Organic Chemistry, Part B, Academic Press, New York, 1973, p. 49.
188   A. Weissberger, W. Schwarze and H. Mainz, Justus Liebigs Ann. Chem., 481 (1930) 68.
189   B.A. Marshall and W.A. Waters, J. Chem. Soc. London, (1961) 1579.
190   S.V. Singh, O.C. Saxena and M.P. Singh, J. Amer. Chem. Soc., 92 (1970) 537.
191   N. Nath and M.P. Singh, J. Phys. Chem., 69 (1965) 2038.
192   J. Hill and A. McAuley, J. Chem. Soc. A, (1968) 1169.
193   C.K. Jørgenson, Absorption Spectra and Chemical Bonding, Pergamon, Oxford, 1962, Chap. 9.
194   J. Hill, A. McAuley and W.F. Pickering, Chem. Commun., (1967) 573.
195   B. Krishna and K.C. Tewari, J. Chem. Soc. London, (1961) 3097.
196   A. McAuley, J. Chem. Soc. London (1965) 4054.
197   C.F. Wells and C. Barnes, Trans. Faraday Soc., 67 (1971) 3297.
198   J.R. Jones, W.A. Waters and J.S. Littler, J. Chem. Soc. London, (1961) 630.
199   J.H. Baxendale, H.R. Hardy and L.H. Sutcliffe, Trans. Faraday Soc., 47 (1951) 963.
200   C.F. Wells and L.V. Kuritsyn, J. Chem. Soc. A, (1969) 2930.
201   C.F. Wells and L.V. Kuritsyn, J. Chem. Soc. A, (1970) 676.
202   C.F. Wells and L.V. Kuritsyn, J. Chem. Soc. A, (1969) 2575.
203   G. Davies and K. Kustin, Trans. Faraday Soc., 65 (1969) 1630.
204   C.F. Wells and L.V. Kuritsyn, J. Chem. Soc, A, (1970) 1372.
205   K. Reinschmiedt, J.C. Sullivan and M. Woods, Inorg. Chem., 12 (1973) 1639.
206a   E. Mentasti and E. Pelizzetti, J. Chem. Soc. Dalton, (1973) 2605.
206b   E. Mentasti, E. Pelizzetti and G. Saini, J. Chem. Soc. Dalton, (1973) 2609.
207   R. Cecil and J.S. Littler, J. Chem. Soc. B, (1968) 1420.
208   R.G. Sandberg, J.J. Auborn, E.M. Eyring and K.O. Watkins, Inorg. Chem., 11 (1972) 1952.
209   T.A. Cooper and W.A. Waters, J. Chem. Soc. B, (1967) 455.

210   P.J. Andrulis and M.J.S. Dewar, J. Amer. Chem. Soc., 88 (1966) 5483.

211a A. McAuley and U.D. Gomwalk, J. Chem. Soc. A, (1969) 977.

211b A. McAuley and R. Shanker, J. Chem. Soc. Dalton, (1973) 2321.

212   U.D. Gomwalk and A. McAuley, J. Chem. Soc. A, (1968) 2948.

213   G. Davies, Inorg. Chem., 11 (1972) 2488.

214   M.C. Agrawal and S.P. Mushran, J. Phys. Chem., 72 (1968) 1497.

215   J. Hill and A. McAuley, J. Chem. Soc. A, (1968) 156.

216   D.K. Lavallee, J.C. Sullivan and E. Deutsch, Inorg. Chem., 12 (1973) 1440.

217   W.F. Pickering and A. McAuley, J. Chem. Soc. A, (1968) 1173.

218   K.J. Ellis and A. McAuley, J. Chem. Soc. Dalton, (1973) 1533.

219   J.J. Bohning and K. Weiss, J. Amer. Chem. Soc., 82 (1960) 4724.

220   R.C. Kapoor, O.P. Kachhwaha and B.P. Sinha, J. Phys. Chem., 73 (1969) 1627.

Chapter 3

OXIDATIONS BY LEAD(IV), THALLIUM(III), MERCURY(II), AND PALLADIUM(II)

1. GENERAL FEATURES OF THE OXIDANTS

Lead(IV) acetate (lead tetraacetate) is a well-known reagent in preparative organic chemistry* and has been used extensively since its introduction by Dimroth and Schweizer in 1923**. Recent reviews have discussed its applicability as an oxidising and acetoxylating agent to a wide variety of organic compounds. Its competitors as an oxidant include periodic acid with which it shows many similarities, both in mode of reaction and in specificity. The main practical difference is that the latter reagent can be used in aqueous or non-aqueous solutions but the use of Pb(IV) acetate must be confined to a completely, or almost completely, non-aqueous medium. (Addition of water eventually leads to decomposition and precipitation of lead dioxide.) In most cases, the medium chosen is glacial acetic acid; a solution of Pb(IV) acetate in this solvent is colourless and quite stable at room temperature. Furthermore, acetic acid is a good solvent for most organic compounds. Reasonably stable solutions can be made in other media such as benzene and pyridine, particularly if a small amount of acetic acid is present, but water must be rigorously excluded.

---

*Norman et al. [1] have published a series of studies on reactions of lead(IV) acetate with a wide variety of organic compounds. Suggested mechanisms are based essentially on product analysis with complementary evidence, in some cases, from electron spin resonance spectra.
**A heterogeneous suspension of lead dioxide in acetic acid had been used by Meyer in 1911 to oxidise anthracene.

The only other satisfactory medium is an alcoholic one in which Pb(IV) acetate forms a fairly stable brown-coloured solution, showing that alcoholysis has occurred; however, the oxidising power is not impeded.

Lead(IV) acetate is readily prepared by the gradual addition of dry red lead oxide to a mixture of glacial acetic acid and acetic anhydride at 40—60°C, the latter reagent being added in calculated amount to remove any water present initially and also that formed by the reaction

$$Pb_3O_4 + 8\,CH_3COOH = Pb(OOCCH_3)_4 + 2\,Pb(OOCCH_3)_2 + 4\,H_2O$$

On cooling, the compound separates out as colourless crystals which, after recrystallisation, can be kept under vacuum or stored moistened with solvent.

Solutions of Pb(IV) acetate in acetic acid show strong absorption in the ultraviolet (log $\epsilon \sim 2.0$ at 300 nm) and this affords a more sensitive alternative to iodometric titrations for monitoring the reagent in kinetic studies [2]. That Pb(IV) acetate is a powerful oxidant is illustrated by its high oxidation potential, at least 1.4 V positive in acetic acid [3].

A little work has been done on the state of Pb(IV) acetate in acetic acid solution. Conductivity measurements show a behaviour to be expected from a non-electrolyte. Also its solubility, 0.07 M in anhydrous acetic acid at 25°C, is decreased in the presence of sodium acetate, a result which might be interpreted on the grounds of a "salting-out" of a non-electrolyte. By way of contrast, the solubility of Pb(II) acetate in acetic acid, 3.3 M at 25°C, is increased by the addition of sodium acetate; aceto-plumbates have been isolated and, in general, conductivity measurements confirm that Pb(II) acetate behaves as a typical weak electrolyte in acetic acid. An important contribution to the knowledge of the lead species present in solution has been made by Evans et al. [4] who made a thorough study of isotopic exchange reactions in acetic acid and acetic anhydride. They observed that rapid exchange of acetate groups took place between Pb(IV) and Pb(II) acetates and the solvent acetic acid, and also between Pb(IV) acetate and acetic anhydride. To account for the exchange between Pb(II) acetate and acetic acid, it was suggested that slight ionisation occurs

$$Pb(OOCCH_3)_2 + CH_3COOH \rightleftharpoons Pb(OOCCH_3)_3^- + H^+$$

with the possibility of association with a solvent molecule

$$Pb(OOCCH_3)_2 + CH_3COOH \rightleftharpoons HPb(OOCCH_3)_3$$

The rapid exchange between Pb(IV) acetate and acetic acid may occur via the formation of similar uncharged complexes. Exchange of radio-lead between Pb(IV) and Pb(II) acetates in acetic acid had been previously reported but Evans et al. have shown this to be spurious: no exchange occurs between the acetates over a period of four hours at 80°C. It is interesting that the observed extinction coefficient for Pb(IV) acetate in acetic acid is found to vary slightly with temperature which suggests a solvation equilibrium has been disturbed. Such an equilibrium is no doubt responsible for the remarkable stability of Pb(IV) in acetic acid compared with other solvents. Ion migration studies have indicated the existence of anionic species of Pb(IV), viz. $Pb(OOCCH_3)_5^-$ and/or $Pb(OOCCH_3)_6^{2-}$, in the presence of acetate ions although, from the lack of spectral changes, the extent of complexing would appear slight.

Thallium(III) is a fairly moderate oxidant which has been little exploited until recently. Kinetic studies, few in number, make use of the perchlorate and acetate which can be prepared by electrolytic oxidation of Tl(I) or directly from thallium(III) oxide or hydroxide. The acetate (now available commercially) is soluble in methanol as well as in acetic acid. Chemically, Tl(III) can be determined by the addition of excess Fe(II) followed by back-titration with Ce(IV) using ferroin, or iodometrically. Alternatively, it can be analysed (and monitored) by its ultraviolet absorption (peak at 242 nm in chloride media [5], 212 nm in perchlorate solution [6]), there being little if any interference from thallium(I). It should be noted that thallium(III) compounds are extremely toxic.

2. REACTIONS WITH INORGANIC SPECIES

(a) Metal ions

Lead(IV) acetate will oxidise cobalt(II) [7] and cerium(III) [8] in acetic acid

$$Pb(IV) + 2 Co(II) = Pb(II) + 2 Co(III)$$

$$Pb(IV) + 2 Ce(III) = Pb(II) + 2 Ce(IV)$$

In the first case, there are kinetic grounds for postulating, as part of the mechanistic scheme, a two-equivalent reaction producing cobalt(IV)

$$Pb(IV) + Co(II) \rightleftharpoons Pb(II) + Co(IV)$$

Of the two reaction products, Pb(II) and Co(III), only Pb(II) has a retarding influence on the rate. The existence of ionic species of reactants is inferred by the increase in rate brought about by the addition of sodium acetate, Pb(IV) reacting as $Pb(OAc)_5^-$ and/or $Pb(OAc)_6^{2-}$.

A number of kinetic studies have been made of reactions between thallium(III) and other metal ions.

$$Tl(III) + 2V(III) = Tl(I) + 2V(IV) \quad (ref.\ 9)$$
$$Tl(III) + 2V(IV) = Tl(I) + 2V(V) \quad (ref.10)$$
$$Tl(III) + 2Fe(II) = Tl(I) + 2Fe(III) \quad (ref.11)$$
$$Tl(III) + Hg(I)_2 = Tl(I) + 2Hg(II) \quad (ref.12)$$
$$Tl(III) + U(IV) = Tl(I) + U(VI) \quad (ref.13)$$
$$Tl(III) + 2Os(bipy)_3^{2+} = Tl(I) + 2Os(bipy)_3^{3+} \quad (ref.14)$$

Aqueous perchlorate solutions were the choice for reaction media except in the case of the U(IV) reaction where a mixed methanol solvent was used. In the majority of these systems there is considerable evidence for the participation of an intermediate oxidation state of thallium(II). For example, in the oxidation of Fe(II), addition of Fe(III) retards the reaction but added Tl(I) has no significant effect. Thus the mechanism would appear to be one involving two successive, one-equivalent reactions

$$Tl(III) + Fe(II) \underset{k_{-1}}{\overset{k_1}{\rightleftharpoons}} Tl(II) + Fe(III)$$

$$Tl(II) + Fe(II) \overset{k_2}{\longrightarrow} Tl(I) + Fe(III)$$

in which Fe(III) and Fe(II) compete for the Tl(II) intermediate. The rate law

$$-\frac{d[Fe(II)]}{dt} = \frac{2k_1k_2[Fe(II)]^2[Tl(III)]}{k_2[Fe(II)] + k_{-1}[Fe(III)]}$$

can be derived assuming a stationary-state concentration for Tl(II) and is in agreement with the kinetic data. Increase in acidity decreases the rate of reaction and Tl(III) would seem to react via the hydrolysed species $Tl(OH)^{2+}$ and $Tl(OH)_2^+$. Sulphate catalyses the reaction but chloride has an inhibiting effect. This is indicative of the formation of complexes of varying reactivities.

The reduction of mercury(II) to mercury(I) represents an overall two-equivalent change since mercury(I) is dimeric. This reduction is brought about by reducing metal ions like V(III) [10] and Cr(II) [15], e.g.

$$2\,Hg(II) \;+\; 2\,Cr(II) \;=\; Hg(I)_2 \;+\; 2\,Cr(III)$$

The mechanism invoked for this last reaction is

$$
\begin{array}{lll}
Hg(II) + Cr(II) \longrightarrow Hg(I) + Cr(III) & & \text{slow} \\
Hg(I) + Cr(II) \longrightarrow Hg(0) + Cr(III) & & \text{rapid} \\
Hg(0) + Hg(II) \longrightarrow Hg(I)_2 & & \text{rapid}
\end{array}
$$

Hg(II) oxidations are catalysed by chloride ions as a result of the formation of the complex $HgCl^+$.

(b) Non-metallic species

Both mercury(I) and mercury(II) oxidise hydrogen, the reactions showing simple second-order kinetics [16]. Hydrido intermediates, e.g. HgH^+, are unlikely for kinetic and energetic reasons and also on the grounds that the solubility of mercury in water is independent of the concentration of hydrogen ions. Instead, direct two-electron transfers would seem to be involved

$$
\begin{array}{ll}
Hg_2^{2+} + H_2 \longrightarrow 2\,Hg\,(or\ Hg_2) + 2\,H^+ \\
Hg^{2+} + H_2 \longrightarrow Hg + 2\,H^+
\end{array}
$$

Hydrogen is not readily oxidised by thallium(III) except in the presence of a Cu(II) catalyst [17a, 17b]. The uncatalysed reaction represented by

$$
\begin{array}{ll}
H_2 \longrightarrow H^- + H^+ \\
H^- + Tl^{3+} \longrightarrow H^+ + Tl^+
\end{array}
$$

is not favoured since the calculated activation energy for the heterolytic splitting of hydrogen is high ($\geqslant 146$ kJ mol^{-1}) as compared with the overall value (109 kJ mol^{-1}) for the catalysed route

$$Cu^{2+} + H_2 \longrightarrow CuH^+ + H^+ \qquad \text{slow}$$

$$CuH^+ + Tl^{3+} \longrightarrow Cu^{2+} + H^+ + Tl^+ \qquad \text{rapid}$$

Thus reaction is facilitated by the catalyst stabilising the H$^-$ intermediate.

Although immune to attack by mercury(I), carbon monoxide is oxidised by mercury(II)

$$2\,Hg^{2+} + CO + H_2O = Hg_2^{2+} + CO_2 + 2\,H^+$$

through a mechanism known as "hydroxymercuration", i.e. the simultaneous coordination of the metal ion and a hydroxide group to the carbon atom of the substrate [18].

$$Hg^{2+} + CO + H_2O \xrightarrow{\text{slow}} \left[Hg-\overset{\overset{O}{\|}}{C}-OH \right]^+ + H^+ \longrightarrow Hg + CO_2 + 2\,H^+$$

$$Hg + Hg^{2+} \rightleftharpoons Hg_2^{2+}$$

Alternatively, this can be looked upon as the insertion of CO between Hg^{2+} and a coordinated water molecule.

$$-Hg^{2+}OH_2 + CO \longrightarrow \left[-Hg-\overset{\overset{O}{\|}}{C}-OH \right]^+ + H^+$$

Evidence for the hydroxymercuration step is the isolation, from methanol solutions, of stable compounds of the type XHg—CO—OCH$_3$ where X is acetate or chloride. Hydroxymercuration occurs also in the oxidation of olefins (see p. 123).

$$Hg^{2+} + RCH=CH_2 + H_2O \longrightarrow [HgCH_2CHROH]^+ + H^+$$

Thallium(III) is not capable of oxidising carbon monoxide.

Thallium(III) oxidises hypophosphite in aqueous perchloric acid solutions [6].

$$Tl(III) + O=\overset{\overset{H}{|}}{\underset{\underset{H}{|}}{P}}-OH + H_2O = Tl(I) + O=\overset{\overset{H}{|}}{\underset{\underset{OH}{|}}{P}}-OH + 2H^+$$

Unlike the oxidation of Hg(I), U(IV), Os(bipy)$_3^{2+}$, and formic acid, there is no rate dependence on hydrogen ion concentration; thus it would appear that $TlOH^{2+}$ is unimportant kinetically. The evidence suggests that prior complex formation occurs with the substrate

$$Tl(III) + H_3PO_2 \rightleftharpoons Tl(H_3PO_2)^{3+}$$

followed by decomposition of the complex directly by

$$Tl(H_3PO_2)^{3+} + H_2O \longrightarrow Tl(I) + H_3PO_3 + 2H^+$$

or indirectly by

$$Tl(H_3PO_2)^{3+} \longrightarrow Tl(I) + H_2PO_2^+ + H^+$$

and

$$H_2PO_2^+ + H_2O \longrightarrow H_3PO_3 + H^+ \qquad rapid$$

However, the possibility of reaction via a Tl(II) intermediate cannot be ruled out, viz.

$$Tl(H_3PO_2)^{3+} \longrightarrow Tl(II) + H_2PO_2 + H^+$$

$$H_2PO_2 + Tl(II) + H_2O \longrightarrow Tl(I) + H_3PO_3 + H^+$$

Chloride ion is an efficient catalyst of this reaction because of the greater reactivity of the Tl(III)—chloride complex.

Phosphorous acid, H_3PO_3, is oxidised further by thallium(III) at higher temperatures of 60—70°C.

$$Tl(III) + H_3PO_3 + H_2O = Tl(I) + H_3PO_4 + 2 H^+$$

It is likely that Tl(III) undergoes complexation with the substrate but, unlike the hypophosphite system, there is an inverse dependence on hydrogen ion concentration [19].

Mercuric chloride oxidises hypophosphorous acid to phosphorous acid and the reaction is reported to have a first-order dependence on substrate concentration but a zero-order dependence on Hg(II) concentration [20]. This, together with the observation that Cu(II) and iodine react at the same rate as Hg(II), points to a slow and rate-determining tautomeric equilibrium (analogous to that in a keto—enol system) followed by a rapid oxidation step.

3. OXIDATION OF OLEFINS

Olefins are oxidised by thallium(III) in sulphuric, nitric, and perchloric acid solutions to a mixture of glycols and carbonyl compounds. In the case of ethylene, the products are acetaldehyde and ethanediol. Henry [21] has shown that the rate is first-order in both ethylene and Tl(III) concentrations, measurements being made on the uptake of ethylene at a constant pressure. The formation of a π-bonded complex is proposed

$$TI^{3+} + CH_2=CH_2 \; \underset{}{\overset{k_1}{\rightleftharpoons}} \; \begin{array}{c} H_2C \\ || \\ H_2C \end{array} \! TI^{3+}$$

which reacts with water to produce a second intermediate

$$\begin{array}{c} H_2C \\ || \\ H_2C \end{array} \! TI^{3+} + H_2O \; \underset{}{\overset{k_2}{\rightleftharpoons}} \; {}^{2+}TICH_2CH_2OH + H^+$$

This is followed by two fast competitive reactions to give either glycol or aldehyde as products

$$^{2+}TICH_2CH_2OH \underset{}{\overset{H_2O}{\longrightarrow}} \begin{cases} TI^+ + (CH_2OH)_2 + H^+ \\ \\ TI^+ + CH_3CHO + H^+ \end{cases}$$

The rate does not depend on the acid concentration; k_1 or k_2 could be rate-determining although, on balance, k_2 is to be preferred since formation of π complexes is usually rapid. Further insight into the mechanism has been provided by a study of the effect of olefin structure on the rate and on the distribution of products. The kinetics of oxidation of propene and the four butenes are identical to those for ethylene but the rates vary over a considerable range. In 0.25 M perchloric acid solutions, the relative rates for ethylene, trans-2-butene, cis-2-butene, 1-butene, propene, and isobutene are 1, 13.6, 58, 162, 167, and $\sim 2 \times 10^5$, respectively, at 25°C [22]. In aqueous acetic acid, because of strong complexing by acetate groups, thallium(III), as $TI(OAc)_3$ or $TI(OAc)_4^-$, is rather unreactive towards olefins although its reactivity is enhanced in the presence of strong acids, an effect presumably caused by removal of coordinated acetate groups by protons [22]. Indeed, it is likely that the reactive species of Tl(III) in this medium is $TI(OAc)_2^+$. At low acetate concentrations, oxidation by TI^{3+} and $TI(OAc)^{2+}$ may become important.

Mercuric chloride in nearly neutral solution reacts with ethylene to form a 1:1 adduct [23a, 23b]

$$H_2C=CH_2 + HgCl_2 + H_2O = ClHgCH_2CH_2OH + HCl$$

The system has an equilibrium constant of ~3 at 25°C but can be taken to completion by the addition of alkali [23b]. Nuclear magnetic resonance spectra have revealed that the carbon—mercury bond is of the σ type. Mercuric salts of strong oxy-acids (e.g. sulphate and trifluoroacetate) oxidise propene to give acraldehyde as the primary product [24]

$$CH_2=CHCH_3 + 4HgSO_4 + H_2O = CH_2=CHCHO + 2Hg_2SO_4 + 2H_2SO_4$$

Again, there is clear evidence that a 1:1 adduct acts as an intermediate, having, in this case, the structure $CH_3CH(OH)CH_2.Hg^+X^-$. In fact, the adduct is stable enough for its reaction with Hg(II) to have been studied separately and it appears that this is the rate-controlling stage in the overall oxidation of olefin. As polymerisation of acraldehyde does not occur, it is apparent that free radicals are not involved in the reaction. There is evidence to suggest that the inter-mediate may have an essentially symmetrical structure of the type

since oxidation of propene labelled with carbon-13 (as $CH_3-CH=\overset{*}{C}H_2$) gives $\overset{*}{C}H_2=CHCHO$ and $CH_2=CH\overset{*}{C}HO$ in equal proportions [25]. Comparison with olefin oxidation by thallium(III) shows similar patterns of behaviour as regards formation of adducts and influence of olefin structure on reactivity. This is to be expected on the grounds that Tl(III) is isoelectronic with the Hg(II) ion. However, one pronounced difference is in the type of product formed; Tl(III) gives glycols along with saturated aldehydes or ketones whereas Hg(II) produces only unsaturated aldehydes or ketones.

Ethylene is oxidised by aqueous palladium(II) chloride to acetaldehyde [26a—26c]

$$C_2H_4 + PdCl_2 + H_2O = CH_3CHO + Pd + 2HCl$$

The uptake of ethylene gas occurs in two stages. Firstly, a rapid reaction takes place where the rate of uptake of gas decreases as the chloride ion increases but is independent of acidity changes. This is attributed to the formation of a π complex

$$PdCl_4^{2-} + C_2H_4 \rightleftharpoons (PdCl_3C_2H_4)^- + Cl^- \qquad K$$

which is hydrolysed by

$$(PdCl_3C_2H_4)^- + H_2O \rightleftharpoons (PdCl_2(H_2O)C_2H_4) + Cl^-$$

where K has a value of 17 at 25°C in perchlorate media. Secondly, a slow reaction, first-order in Pd(II), takes over whose rate is reduced by chloride ions and by hydrogen ions

$$-\frac{d[C_2H_4]}{dt} = \frac{k[PdCl_3C_2H_4^-]}{[H^+][Cl^-]}$$

The inverse hydrogen ion dependence is likely to arise from the formation of a π-bonded hydroxo species

$$(PdCl_2(H_2O)C_2H_4) + H_2O \rightleftharpoons (PdCl_2(OH)C_2H_4)^- + H_3O^+$$

which decomposes in a slow, rate-controlling step to give a σ-bonded intermediate which in turn quickly breaks down yielding products.

$$(PdCl_2(OH)C_2H_4)^- \longrightarrow ClPdCH_2CH_2OH + Cl^- \qquad \text{slow}$$

$$ClPdCH_2CH_2OH \longrightarrow HCl + Pd + CH_3CHO \qquad \text{rapid}$$

In acetic acid, Pd(II) oxidises ethylene to vinyl acetate

$$Pd^{2+} + C_2H_4 + 2\,OAc^- = Pd + CH_2{=}CHOAc + HOAc$$

and other olefins to vinyl and allylic acetates. However, a combination of Pd(II) and Cu(II) in acetic acid can be made to produce acetate esters of chloro alcohols and glycols of the type $AcOCH_2CH_2Cl$, $AcOCH_2CH_2OAc$, and $AcOCH_2CH_2OH$ [27a, 27b]. (By itself, Cu(II) is incapable of reaction with olefins.) The reaction is seen in terms of the formation of an adduct with Pd(II) which then reacts with Cu(II) chloride to produce saturated esters, Cu(II) being reduced to Cu(I) in the process, e.g.

$$C_2H_4 + PdCl_2 + OAc^- \xrightarrow{\text{HOAc}} ClPdCH_2CH_2OAc + Cl^-$$

$$ClPdCH_2CH_2OAc \xrightarrow{2\,CuCl_2} \begin{cases} PdCl_2 + ClCH_2CH_2OAc + 2\,CuCl \\ CH_2{=}CHOAc + HPdCl \end{cases}$$

A competitive path is available in which decomposition of the adduct takes place by elimination of HPdCl (i.e. Pd(0)) to give vinyl acetate, the reaction conditions deciding which path predominates.

Palladium(II) in conjunction with Cu(II) is made use of in the important and commercially viable reaction known as the Wacker process (sometimes referred to as the Smidt reaction) in which acetaldehyde is produced from ethylene [28a, 28b]. This utilises aerobic oxidation to reoxidise the Cu(I) produced in the reduction by palladium metal of Cu(II) and the basic steps are

$$Pd^{II}Cl_4{}^{2-} + C_2H_4 + H_2O = Pd + CH_3CHO + 2H^+ + 4Cl^-$$

$$Pd + 2Cu(II) + 4Cl^- = Pd^{II}Cl_4{}^{2-} + 2Cu(I)$$

$$2Cu(I) + 2H^+ + \tfrac{1}{2}O_2 = 2Cu(II) + H_2O$$

The net reaction

$$C_2H_4 + \tfrac{1}{2}O_2 = CH_3CHO$$

shows that only ethylene and oxygen are consumed, palladium(II) and Cu(II) being continually regenerated.

Rhodium(III) chloride is effective as an oxidant for ethylene in a solvent of dimethylacetamide and produces a mixture of acetaldehyde, but-1-ene, and but-2-ene [29]. A possible scheme is

$$Rh^{III}Cl_n \rightleftharpoons Rh^{III}Cl_{n-1} + Cl^-$$

$$Rh^{III}Cl_{n-1} + C_2H_4 \rightleftharpoons Rh^{III}Cl_{n-1}(C_2H_4)$$

$$Rh^{III}Cl_{n-1}(C_2H_4) + H_2O \rightleftharpoons Rh^ICl_{n-1} + 2H^+ + CH_3CHO \quad \text{rapid}$$

$$Rh^ICl_{n-1} + C_2H_4 \longrightarrow Rh(I) \text{ complex} \quad \text{rapid}$$

where the water utilised in the third step originates in the coordination sphere of Rh(III).

As an extension of the work of Henry with simple olefins, Byrd and Halpern [30] have investigated the Tl(III) oxidation of a number of alkenols in perchloric acid media. In some cases, e.g. with 1-propenol-3-ol, direct spectral evidence is adduced for the presence of intermediate species, and the results are compatible with an oxythallation—dethallation mechanism of the type

$$Tl^{3+} + CH_2=CHR + H_2O \longrightarrow [TlCH_2CH(R)OH]^{2+} + H^+$$

$$[TlCH_2CH(R)OH]^{2+} \longrightarrow Tl^+ + products$$

Additional support for this interpretation is provided by the kinetic behaviour shown by cycloalkenes and methylenecycloalkanes [31].

The kinetics of oxidation of a series of substituted styrenes [32] and arylcyclopropanes [33] by Tl(III) acetate in acetic acid have been reported by Ouellette and co-workers; at 50°C, the reactivity order in both systems is p-CH$_3$O $> p$-CH$_3 >$ m-CH$_3 >$ H $> p$-Cl $> m$-Cl. As reaction progresses, the Tl(I) product removes Tl(III) to form a relatively unreactive double salt

$$Tl(OAc)_3 + TlOAc \rightleftharpoons Tl_2(OAc)_4$$

bringing about a progressive decrease in the rate of oxidation of the substrates. The cleavage of phenylcyclopropane by mercury(II) acetate is approximately ten times slower than cleavage by thallium(III) acetate (at 25°C).

4. OXIDATION OF ALCOHOLS AND ETHERS

Mercury(II) perchlorate, like bromine, is relatively inert to methanol and attack on ethanol is only very slow. However, secondary alcohols are readily oxidised, cyclohexanol being more readily attacked than isopropanol. This sensitivity to the degree of substitution of the carbon atom at the site of oxidation is not met with in the case of chromium(VI) oxida-

tions which occur via an ester intermediate (p. 179). Since the
rates of oxidation of diisopropyl ether and secondary alcohols
by bromine and Hg(II) are comparable, it appears that the
O—H bond of the alcohol and the O—R bond of the ether
remain intact in the transition stage and the mechanisms are
accordingly hydride ion transfers [34], viz.

where R is H or an alkyl group.

5. OXIDATION OF HYDROPEROXIDES

Oxidation of *tert.*-butyl hydroperoxide by Pb(IV) acetate
has been studied in detail by Benson and Sutcliffe [35]. The
stoichiometry of the reaction in anhydrous acetic acid is
essentially 2 :1 showing that Pb(IV) or species derived from it
do not function as catalysts. The products of reaction are
oxygen, *tert.*-butyl alcohol, acetone, formaldehyde, and di-
tert.-butyl peroxide. Kinetically, the reaction is first-order

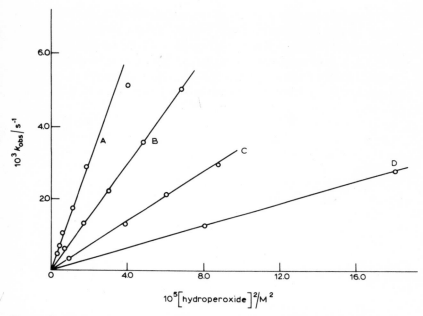

Fig. 1. Oxidation of *tert.*-butyl hydroperoxide by lead(IV) acetate. Approximate second-order dependence of reaction rate on peroxide concentration at A, 37.85°C; B, 31.03°C; C, 23.75°C; and D, 17.65°C. (From Benson and Sutcliffe [35], by courtesy of The Chemical Society.)

in Pb(IV) and approximately second-order in hydroperoxide (Fig. 1), rates being followed by monitoring the disappearance of Pb(IV) spectrophotometrically at 320 nm. The non-catalytic nature of the reaction is demonstrated further by the failure of Pb(II) to affect the rate even at high concentrations. Presumably, the redox potential of the Pb(IV)—Pb(II) couple is too high to permit the oxidising entities, resulting from peroxide breakdown, to convert Pb(II) to Pb(IV).

The approximately second-order dependence on peroxide

can be explained in two ways: either the hydroperoxide reacts via its dimer or it forms a 1:1 complex with Pb(IV). Although hydroperoxides can exist in polymeric forms in non-polar solvents, it is improbable that such species are present in acetic acid since the latter might be expected to associate with peroxide molecules. Moreover, the reaction preserves its over-all third-order character even in the presence of water when hydrogen bonding between peroxide molecules would be much reduced. Thus it is likely that the rate-determining step is that between Pb(IV) and a Pb(IV)—peroxide complex. A subsidiary direct reaction between Pb(IV) and peroxide could account for the observation that the reaction is slightly less than second-order in the latter.

$$\text{Pb(IV)} + \text{ROOH} \rightleftharpoons \text{Pb(IV)·ROOH} \qquad K_1$$

$$\text{Pb(IV)·ROOH} + \text{ROOH} \xrightarrow{k_1}$$

$$\text{Pb(IV)} + \text{ROOH} \xrightarrow{k_2}$$

The rate law expressed in corresponding form is

$$-\frac{d[\text{Pb(IV)}]}{dt} = k_1 K_1 [\text{Pb(IV)}][\text{ROOH}]^2 + k_2 [\text{Pb(IV)}][\text{ROOH}]$$

The slight reduction in peroxide order with increase in tem-perature indicates that a negative enthalpy change is associated with K_1.

Sodium acetate brings about a marked acceleration, the rate being directly proportional to the concentration (i.e. to the *square* of the acetate ion concentration since sodium acetate functions as a weak electrolyte in acetic acid) as is found for the thermal decomposition of Pb(IV) (p. 137). Also, there is a reduction in the peroxide dependence from 2 to 1.5. It is clear that the reactive species of Pb(IV) is Pb(OAc)_6^{2-} and also that this acetate complex can undergo still further complexation with peroxide. The reaction (both with and without acetate) is unaffected by addition of sodium perchlorate. This inert salt has one of the largest dissociation constants in acetic acid, is non-complexing and affords a means whereby ionic salt effects can be studied. The negative result is noteworthy in that it suggests that not more than one reactant can be ionic.

Extrapolation of rate data to zero time gives no indication of a peroxide complex. However, it is possible to confirm the existence of such a complex by adding a complexing agent more powerful than peroxide. Well-known complexing agents like EDTA are unsuitable because of low solubility but alcohols form complexes which are sufficiently stable to be useful. The order of reaction is reduced in the presence of ethanol, the rate being directly proportional to the concentration of alcohol and it is likely that the latter partially supplants peroxide in the initial complex, forming a highly reactive 1:1 Pb(IV)—alcohol species, i.e. the step Pb(IV).ROOH + ROOH → is now in competition with Pb(IV).R'OH + ROOH →. In a pure alcohol solvent, the latter reaction takes over completely and the kinetics become first-order in peroxide.

Product analysis for the *tert.*-butyl hydroperoxide + Pb(IV) acetate reaction shows a distribution of products similar to those found by Dean and Skirrow [36] for the catalysed decomposition of the same hydroperoxide by cobalt(II) acetate. After the initial rate-determining steps

$$Pb(IV)\cdot ROOH + ROOH \longrightarrow Pb(III)\cdot ROOH + RO_2^{\cdot} + H^+$$

$$(or \ Pb(IV)\cdot ROOH + ROOH \longrightarrow Pb(II) + 2 RO_2^{\cdot} + 2 H^+)$$

$$Pb(IV) + ROOH \longrightarrow Pb(III) + RO_2^{\cdot} + H^+$$

which generate peroxy radicals, the subsequent stages are postulated as

$$RO_2\cdot + RO_2\cdot \longrightarrow ROOR + O_2$$

$$RO_2\cdot + RO_2\cdot \longrightarrow 2\,RO\cdot + O_2$$

$$RO_2\cdot + ROOH \longrightarrow ROH + O_2 + RO\cdot$$

$$RO\cdot + ROOH \longrightarrow ROH + RO_2\cdot$$

where ROOR represents di-*tert.*-butyl peroxide and ROH is *tert.*-butyl alcohol. Formaldehyde and acetone are accounted for by the combination of RO$\cdot$ and RO$_2^\cdot$ radicals

$$(CH_3)_3CO\cdot + (CH_3)_3CO_2\cdot \longrightarrow (CH_3)_3COOCH_3 + (CH_3)_2CO$$

then

$$(CH_3)_3COOCH_3 \longrightarrow (CH_3)_3COH + H_2CO$$

The decomposition of *tert.*-butyl methyl peroxide, $(CH_3)_3COOCH_3$, has not been studied but the analogous peroxides, methyl hydroperoxide and dimethyl peroxide, are reported to give formaldehyde as part of their breakdown products.

Homolytic fission of a hydroperoxide molecule can be brought about by other metal ions, e.g. by iron(II) and cobalt(II) [37], and these types of reaction are of particular importance in autoxidation processes. There is ample evidence to indicate that peroxide complexes are as significant here as in the Pb(IV) case. Reaction with Fe(II) is a chain of short kinetic length in which Fe(II) is oxidised irreversibly to Fe(III). But cobalt(II) initiates a chain decomposition of considerable length since Co(III), the oxidised form, can react further with hydroperoxide, being reduced back to Co(II) again, and thus the metal ion behaves as a redox catalyst, i.e.

$$Co(II)\cdot ROOH \longrightarrow RO\cdot + OH^- + Co(III)$$

then

$$Co(III) + ROOH \longrightarrow RO_2^{\cdot} + H^+ + Co(II)$$

and so on. The proof of the existence of radicals in these decompositions comes from a study of the reaction conducted in the presence of olefins (e.g. butadiene, cyclohexene, oct-1-ene, and isoprene) when identifiable products were isolated, resulting from the attack of $RO^{\cdot}$ and $RO_2^{\cdot}$ radicals on the olefin.

Evidence for radical participation in peroxide systems arises from observations made by Stannett and Mesrobian [38] on the thermal decomposition of cumene and tert.-butyl hydroperoxides in a wide variety of solvents over the temperature range 50—100°C. That the reaction involves free radical intermediates is demonstrated by the polymerisation produced in monomers and also from the fact that the kinetic order varies in the presence of styrene monomer. It is interesting that the photochemical decomposition of these hydroperoxides in carbon tetrachloride, hexane, and dioxan solutions would seem to take place by the intervention of $RO^{\cdot}$ and $RO_2^{\cdot}$ radicals also but diperoxide is not found as a product [39].

6. OXIDATION OF GLYCOLS

In 1931, Criegee showed that lead(IV) acetate is an efficient and rapid oxidising agent for vicinal glycols, converting the $\supset C(OH).C(OH)\subset$ group to two ketone molecules. Arising from kinetic studies of the reaction with a number of glycols, Criegee et al. [40] suggested the mechanism might involve a cyclic intermediate, viz.

$$\begin{array}{l} -\overset{|}{\underset{|}{C}}-OH \\ -\overset{|}{\underset{|}{C}}-OH \end{array} + \; Pb(OOCCH_3)_4 \;\overset{A}{\rightleftharpoons}\; \begin{array}{l} -\overset{|}{\underset{|}{C}}-OPb(OOCCH_3)_3 \\ -\overset{|}{\underset{|}{C}}-OH \end{array} + \; CH_3COOH$$

$$\Big\downarrow \; \text{B (rate-determining)}$$

$$\begin{array}{l} -\overset{|}{C}-O \\ \overset{|}{}\diagdown \\ \overset{|}{}Pb(OOCCH_3)_2 \\ -\overset{|}{C}-O\diagup \end{array} + \; CH_3COOH$$

$$\Big\downarrow \; \text{rapid}$$

$$\begin{array}{l} -\overset{|}{C}=O \\ + \; Pb(OOCCH_3)_2 \\ -\overset{|}{C}=O \end{array}$$

Support for stage A was in the isolation of a yellow product $Pb(OOCCH_3)_2(OH)(OCH_3)$ from the reaction of lead(IV) acetate with methanol containing a small amount of water, and $Pb(OOCCH_3)_3(OCH_3)$ from dry methanol. On the assumption that step B is rate-determining, the observed second-order kinetics were satisfied and the overall mechanism provided a reasonable explanation for the experimental observation that a *cis*-glycol reacts much more readily than its *trans* isomer. However, the cyclic intermediate has never been isolated nor is the mechanism adequate in all cases of glycol scission: some large diol molecules (e.g. dihydrophenanthrenediol) react more rapidly as the *trans* form and substitution of hydrogen or alkyl groups by an aryl group in these glycols can cause a reversal of this effect [41]. Further points worthy of note are:

(1) The rate of glycol cleavage is increased in the presence of acetate, a result indicative of the formation of acetato anionic species of lead(IV) (see p. 130).

(2) Periodic acid and sodium bismuthate bear a formal resemblance to lead(IV) acetate in that their efficiency as glycol cleavers might lie in their ability to form a

$R\overset{O}{\underset{O}{\diamond}}M$ type intermediate with the diol.

(3) Second-order rate constants for oxidation of ethanediol, propane-1,2-diol, and butane-2,3-diol differ considerably (0.420, 3.36, and 45.9 $1 \text{ mol}^{-1} \text{ min}^{-1}$, respectively, at 40°C) [42]. The origin of this marked progression is certainly not forthcoming in terms of the cyclic mechanism.

An alternative cyclic mechanism which has been suggested [41] involves attachment of a lead atom to the *cis*-diol followed by the intramolecular transfer of a proton.

$$\begin{array}{c} -\overset{|}{C}-OH \\ -\overset{|}{C}-OH \\ | \end{array} + Pb(OOCCH_3)_4 \longrightarrow \begin{array}{c} \diagdown\overset{|}{C}\diagup O-Pb(OOCCH_3)_2 \\ \diagdown O \\ -\overset{|}{C}-O-H\cdots O=\overset{|}{C}CH_3 \\ \diagup \end{array} + CH_3COOH$$

$$\downarrow$$

$$\begin{array}{c} -\overset{|}{C}=O \\ -\overset{|}{C}=O \\ | \end{array} + Pb(OOCCH_3)_2 + CH_3COOH$$

Trans-diols, deprived of a cyclic transition state, might then react in an analogous manner by coordination of a hydroxy group to the lead atom followed by proton transfer to a base, e.g. acetate ion.

$$B + \begin{array}{c} -\overset{|}{C}-O-Pb(OOCCH_3)_3 \\ H-O-\overset{|}{C} \\ | \end{array} \longrightarrow BH^+ + \begin{array}{c} -\overset{|}{C}=O \\ O=\overset{|}{C}- \\ | \end{array} + Pb(OOCCH_3)_3^-$$

Although addition of acetate does have a catalytic effect on such oxidations [43], it is unlikely that the slow but measurable reaction in acetic acid solvent can be accommodated on such a scheme because of the absence of significant amounts of free base, but the rapid rates of oxidation observed in pyridine may be attributed to the high basicity of this solvent.

Thallium(III) and mercury(II) would appear to be incapable of oxidising glycols.

7. OXIDATION OF CARBOXYLIC ACIDS

Formic acid is oxidised quantitatively to CO_2 by Tl(III), Hg(II), and Pd(II) in aqueous acid solution. For the first two oxidants, the rate law is of the simple type [44]

$$- \frac{d[HCOOH]}{dt} = \frac{k[HCOOH][Ox]}{[H^+]}$$

For both Tl(III) and Hg(II) the inverse dependence on hydrogen ion concentration probably arises from the ionisation of formic acid to give the easily oxidisable formate anion. A complication is the possibility of complex formation with the substrate

$$M^{n+} + HCOO^- \rightleftharpoons M(HCOO)^{(n-1)+}$$

The rate-controlling step is regarded as a two-equivalent one

$$M(HCOO)^{(n-1)+} \longrightarrow H^+ + CO_2 + M^{(n-2)+}$$

Monobasic carboxylic acids, with the exception of formic acid, are relatively inert to Pb(IV) acetate, being oxidised only at high temperatures. Formic acid is quantitatively oxidised to carbon dioxide at moderate temperatures.

$$HCOOH + Pb(OOCCH_3)_4 = 2 CH_3COOH + Pb(OOCCH_3)_2 + CO_2$$

A formal mechanism for the reaction is

$$HCOOH + Pb(OOCCH_3)_4 \rightleftharpoons CH_3COOH + \begin{array}{c} O \\ \| \\ O=C \\ | \\ H \end{array} \begin{array}{c} O \\ \diagdown \\ Pb(OOCCH_3)_2 \\ | \\ OOCCH_3 \end{array}$$

$$\longrightarrow CO_2 + CH_3COOH + Pb(OOCCH_3)_2$$

At least four independent investigations have been made of the decomposition of Pb(IV) acetate in anhydrous acetic acid at elevated temperatures. However, the detailed mechanism is still unclear. At reflux temperatures of 118°C, the decomposition is heterogeneous in character and it is likely that a simple reaction is being complicated by side effects. Benson et al. [2] have shown that the presence of sodium acetate appears to eliminate the heterogeneous part of the reaction and the decomposition can be studied at much lower temperatures. With a constant excess of sodium acetate, the rate of disappearance of Pb(IV) follows an approximate second-order rate law in the temperature range 50—80°C but increase of the initial Pb(IV) concentration brings about a decrease in the rate. Similar effects have been noted by Levy and Szwarc [45] for the thermal decomposition of diacetyl peroxide in acetic acid, and by Stannett and Mesrobian [38] for the decomposition of cumene hydroperoxide in benzyl alcohol. This result can be explained in terms of the formation and subsequent breakdown of a Pb(IV)—retarder complex. Support for this comes from the observation that concentrated solutions of Pb(IV) acetate and sodium acetate were shown to proceed to a limit of 25% reaction, complete reaction being achieved only on boiling. The kinetics disclose that the rate is directly proportional to the concentration of the sodium acetate concentration (i.e. to the square of the acetate ion concentration). This suggests that the reactive species of Pb(IV) is one containing two acetate ions, i.e. $Pb(OOCCH_3)_6^{2-}$. The reaction mechanism in the presence of acetate can be conceived in terms of the following scheme.

(1) The reactive ion $Pb(OOCCH_3)_6^{2-}$ forms the cation $^+CH_2COO$ via the radical $\cdot CH_2COOH$.

$$Pb^{IV}(OOCCH_3)_6^{2-} + CH_3COOH \longrightarrow Pb^{III}(OOCCH_3)_5^{2-} + CH_3COOH + \cdot CH_2COOH$$

$$Pb^{IV}(OOCCH_3)_6^{2-} + \cdot CH_2COOH \longrightarrow Pb^{III}(OOCCH_3)_6^{3-} + {}^+CH_2COOH$$

(2) The carbonium ion $^+CH_2COOH$ thus formed and a Pb(IV) species join to form a stable intermediate

$$Pb(OOCCH_3)_4 + {}^+CH_2COOH \rightleftharpoons Pb(OOCCH_3)_4({}^+CH_2COOH)$$

or, taking account of solvation factors

$$Pb(OOCCH_3)_4(CH_3COOH) + {}^+CH_2COOH \rightleftharpoons (HOOC\cdot H_2C)Pb(OOCCH_3)_5 + H^+$$

The latter is regarded as the Pb(IV)—retarder complex which only breaks down when the solution is subjected to boiling, the main product of reaction, acetoxyacetic acid, resulting from

$$CH_3COO^- + {}^+CH_2COOH \longrightarrow CH_3COO\cdot CH_2COOH$$

(3) The other product of reaction, succinic acid, probably results from the association of two $\cdot CH_2COOH$ radicals*

$$\cdot CH_2COOH + \cdot CH_2COOH \longrightarrow (CH_2COOH)_2$$

Norman and Poustie [47] have confirmed and extended these kinetic observations showing that as the concentration of Pb(IV) is reduced, the reaction, as well as becoming more rapid, changes over from second- to first-order in type (Fig. 2). For

---

*Succinic acid is formed also as a product in the photolysis of Pb(IV) acetate in acetic acid [46].

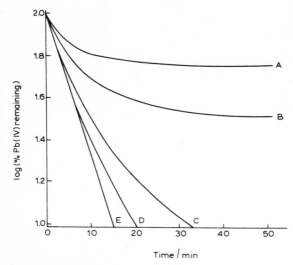

Fig. 2. Decomposition of lead(IV) acetate in acetic acid. Rate of consumption of Pb(IV) in acetic acid containing 1 M sodium acetate at 65°C. Initial concentration of Pb(IV) acetate: A, 10.6×10^{-4} M; B, 7.4×10^{-4} M; C, 5.3×10^{-4} M; D, 3.70×10^{-4} M; and E, 2.11×10^{-4} M. (From Norman and Poustie [47], by courtesy of The Chemical Society.)

the reaction in the presence of acetate, they hold the view that an acetate-catalysed dimerisation is in competition with the catalysed removal of mononuclear Pb(IV) species and that a complex (or a series of complexes) is formed which is relatively stable. In support of this idea, they were able to isolate a binuclear Pb(IV) complex, readily reverting to Pb(IV) acetate when washed with acetic acid, which they speculate might have the structure

$$(AcOH)(AcO)_3\bar{P}b \diagdown \begin{smallmatrix} CH_3 \\ | \\ O-C-O \\ \diagup \quad \overset{+}{} \quad \diagdown \\ O-\overset{+}{C}-O \\ | \\ CH_3 \end{smallmatrix} \diagup \overset{-}{P}b(OAc)_3(HOAc)$$

The first-order path which is favoured at low concentrations is shown to be consistent with a generalised mechanism in which the acetate ion (functioning as a base) abstracts a proton from an acetate group, a process accompanied by the heterolytic fission of two Pb—O bonds to yield acetoxyacetic acid, the main product of reaction

$$B: + \begin{smallmatrix} AcO-Pb(OAc)_2 \\ H \diagdown \quad \diagdown O \\ \diagdown CH_2-CO \end{smallmatrix} \longrightarrow AcO \cdot CH_2CO_2^- + BH^+ + Pb(OAc)_2$$

In the absence of sodium acetate, the decomposition of Pb(IV) acetate in acetic acid at reflux temperature has been shown by Kharasch et al. [48] to include methane and carbon dioxide as products. They advocate a complex free-radical mechanism whose primary stage is

$$Pb(OOCCH_3)_4 + CH_3COOH \longrightarrow Pb(OOCCH_3)_3 + CH_3COOH + \cdot CH_2COOH$$

The generation of acetoxy radicals, $CH_3COO\cdot$, in the initial stage is excluded on the grounds that the thermal decomposition of diacetyl peroxide, a reaction proved to involve acetoxy radicals, gives a product distribution quite different from that encountered in the decomposition of Pb(IV) acetate. If $CH_3COO\cdot$ radicals participate in the subsequent stages, they are thought of as being derived indirectly from $Pb(OOCCH_3)_3$. Carbon dioxide and CH_3 radicals are produced either directly from the latter species or indirectly from the fragmentation of $CH_3COO\cdot$ radicals. Methane is then formed by abstraction of hydrogen from a solvent molecule.

$$Pb(OOCCH_3)_3 \longrightarrow Pb(OOCCH_3)_2 + CH_3COO\cdot$$

$$Pb(OOCCH_3)_3 \longrightarrow Pb(OOCCH_3)_2 + CO_2 + CH_3\dot{}$$

$$CH_3COO\cdot \longrightarrow CO_2 + CH_3\cdot$$

$$CH_3\cdot + CH_3COOH \longrightarrow CH_4 + \cdot CH_2COOH$$

An alternative ionic scheme has been provided by Mosher and Kehr [49] who, on the evidence of product analysis alone, have formulated a mechanism based upon heterolytic fission

$$Pb(OOCCH_3)_4 \longrightarrow Pb(OOCCH_3)_2 + CH_3COO^+ + CH_3COO^-$$

The formation of carbon dioxide and methane are then explained in terms of the steps

$$CH_3COO^+ \longrightarrow CH_3^+ + CO_2$$

$$CH_3^+ + CH_3COOH \longrightarrow CH_4 + CH_3COO^+$$

However, following on the lines of the mechanism for the catalysed reaction, the CH_3COO^+ ion could result from the interaction of $^+CH_2COOH$ with a solvent molecule.

$$^+CH_2COOH + CH_3COOH \longrightarrow CH_3COOH + CH_3COO^+$$

The combination of $^+CH_2COOH$ and CH_3COO^- ions to yield acetoxyacetic acid is not favoured since few acetate ions are available, and thus methane and carbon dioxide predominate as products.

Oxalic acid is oxidised by thallium(III) in aqueous perchlorate solutions

$$Tl^{3+} + H_2C_2O_4 = Tl^+ + 2 CO_2 + 2 H^+$$

in a second-order manner [5]. Strong complex formation occurs and it is likely that the decomposition of an intermediate Tl(III)—oxalate complex, $TlC_2O_4^+$, brings about the rupture of the C—C bond

$$Tl^{3+} + H_2C_2O_4 \rightleftharpoons TlC_2O_4^+ + 2H^+$$

$$TlC_2O_4^+ \longrightarrow Tl^+ + 2CO_2$$

Oxalic acid is also oxidised by Hg(II) chloride but the reaction is very slow unless induced by the addition of small amounts of Mn(II) ions, Cr(VI), or permanganate [50]. Since the oxidation can be brought about equally well by photochemical means, it is likely that a chain mechanism is operative, possibly along the general lines of

$$Ox + H_2C_2O_4 \longrightarrow CO_2 + CO_2^- + 2H^+ + Ox^-$$

$$CO_2^- + HgCl_2 \longrightarrow CO_2 + HgCl + Cl^-$$

$$HgCl + H_2C_2O_4 \longrightarrow Hg + Cl^- + CO_2^- + CO_2 + 2H^+$$

Gold(III) is capable of oxidising oxalic acid, the reaction occurring in two stages

$$AuCl_4^- + H_2C_2O_4 = AuCl_2^- + 2CO_2 + 2HCl$$

followed by a slow reduction of Au(I) to metallic gold. The overall reaction is very much faster under conditions where hydrolysis of chloroaurate(III) occurs [51].

Oxidative cleavage of 2-hydroxycarboxylic acids (e.g. mandelic acid)

$$Pb(OAc)_4 + R_1-\overset{OH}{\underset{R_2}{\overset{|}{C}}}-COOH = Pb(OAc)_2 + R_1-\overset{O}{\overset{\|}{C}}-R_2 + CO_2 + 2HOAc$$

follows simple second-order kinetics and the evidence suggests that a cyclic intermediate is formed [52].

8. OXIDATION OF KETONES

Cyclohexanone is oxidised by Hg(II), Tl(III), MnO_4^-, iodine, and bromine under acid conditions. As the rate is independent of Hg(II) concentration and as all the oxidants react at a similar rate, the rate-determining stage is likely to be an enolisation equilibrium [53a, 53b].

144

Further oxidation to give cyclohexane-1,2-dione occurs under more stringent conditions. It would appear that lead(IV) acetate reacts with ketones in a similar fashion to Hg(II) and Tl(III).

9. OXIDATIVE COUPLING

Reaction of benzene with Pd(II) acetate [54] (or Pd(II) chloride and sodium acetate [55]) in an acetic acid solvent results in the formation of biphenyl.

$$Pd(II) + 2 C_6H_6 = Pd(0) + 2 H^+ + C_6H_5-C_6H_5$$

In the presence of oxygen, the palladium is oxidised back to Pd(II) [56]

$$Pd(0) + 2 H^+ + \tfrac{1}{2}O_2 = Pd(II) + H_2O$$

and, under these conditions, the overall reaction becomes

$$2 C_6H_6 + \tfrac{1}{2}O_2 = C_6H_5-C_6H_5 + H_2O$$

This system is an example of the class of reactions known as oxidative couplings. The rate of formation of biphenyl from benzene in acetic acid solutions is first-order in both substrate and $PdCl_2$ concentrations

$$\frac{d[biphenyl]}{dt} = k[PdCl_2][C_6H_6]$$

and is independent of the concentration of sodium acetate. However, no coupling takes place in the total absence of sodium acetate. The second-order rate law is preserved when

Pd(II) acetate is the oxidant. Substitution of deuterium for hydrogen in the benzene molecule results in a substantial isotope effect. Another feature is the absence of deuteration in the ring when Pd(II) acetate is used in a DOAc solvent. The general indication is that the primary and rate-controlling step in the reaction is the formation of an adduct

$$PdX_2 + C_6H_6 \longrightarrow \text{(adduct)} \qquad \text{slow}$$

which is then converted to a σ-aryl complex as represented by

$$\text{(adduct)} + NaOAc \longrightarrow C_6H_5PdX + NaX + HOAc$$

Decomposition of this complex would seem to involve the initial formation of Pd(I) [56].

$$2 C_6H_5PdX \longrightarrow C_6H_5-C_6H_5 + 2 PdX$$

In the absence of oxygen at temperatures below 60°C, Pd(II) acetate reacts with benzene in an acetic acid—perchloric acid solvent to give, not palladium metal, but a vividly coloured Pd(I) complex, $[Pd(C_6H_6)(H_2O)(ClO_4)]_n$. With chloride ion present, the Pd(I) species disproportionates to Pd(II) and Pd(0). In refluxing acetic acid and in the presence of acetate, Pd(II) brings about the coupling of benzene with ethylene to give styrene and styrene itself couples with benzene to give *trans*-stilbene [57]

$$Pd(OAc)_2 + C_6H_5CH=CH_2 + C_6H_6 = Pd(0) + 2HOAc + C_6H_5CH=CHC_6H_5$$

Pd(II) can be regenerated by the addition of Cu(II) acetate.

A concise account of oxidative coupling reactions has been provided by Stern [58].

146

REFERENCES

1   R.O.C. Norman et al., J. Chem. Soc. London, (1964) 4860; see
    also J.W. Lown, J. Chem. Soc. B, 441 (1966) 644.
2   D. Benson, L.H. Sutcliffe and J. Walkley, J. Amer. Chem. Soc.,
    81 (1959) 4488.
3   A. Berka, V. Dvořák, I. Němec and J. Zýka, J. Electroanal. Chem.,
    4 (1962) 150.
4   E.A. Evans, J.L. Huston and T.N. Norris, J. Amer. Chem. Soc.,
    74 (1952) 4985.
5   L.B. Mønsted, O. Mønsted and G. Nord, Trans. Faraday Soc.,
    66 (1970) 936.
6   K.S. Gupta and Y.K. Gupta, J. Chem. Soc. A, (1970) 256.
7   D. Benson, P.J. Proll, L.H. Sutcliffe and J. Walkley, Discuss.
    Faraday Soc., 29 (1960) 60.
8   D. Benson and L.H. Sutcliffe, Trans. Faraday Soc., 56 (1960) 246.
9   N.A. Daugherty, J. Amer. Chem. Soc., 87 (1965) 5026.
10 W.C.E. Higginson, D.R. Rosseinsky, J.B. Stead and A.G. Sykes,
    Discuss. Faraday Soc., 29 (1960) 49.
11 K.G. Ashurst and W.C.E. Higginson, J. Chem. Soc., (1953) 3044.
12 A.M. Armstrong and J. Halpern, Can. J. Chem., 35 (1957) 1020.
13 A.C. Harkness and J. Halpern, J. Amer. Chem. Soc., 81 (1959)
    3526.
14 D.H. Irvine, J. Chem. Soc., (1957) 1841.
15 J. Doyle and A.G. Sykes, J. Chem. Soc. A, (1968) 215.
16 G.J. Korinek and J. Halpern, J. Phys. Chem., 60 (1956) 285.
17a E. Peters and J. Halpern, J. Phys. Chem., 59 (1955) 793.
17b J. Halpern, E.R. MacGregor and E. Peters, J. Phys. Chem., 60
    (1956) 1455.
18 A.C. Harkness and J. Halpern, J. Amer. Chem. Soc., 83 (1961) 1258.
19 K.S. Gupta and Y.K. Gupta, J. Chem. Soc. A, (1971) 1180.
20 A.D. Mitchell, J. Chem. Soc. London, 119 (1921) 1266.
21 P.M. Henry, J. Amer. Chem. Soc., 87 (1965) 990, 4423.
22 P.M. Henry, J. Amer. Chem. Soc., 88 (1966) 1597.
23a J.C. Chatt, Chem. Rev., 48 (1951) 7.
23b E.R. Allen, J. Cartlidge, M.M. Taylor and C.F.H. Tipper, J. Phys.
    Chem., 63 (1959) 1437.
24 B.C. Fielding and H.L. Roberts, J. Chem. Soc. A, (1966) 1627.
25 J.C. Strini and J. Metzger, Bull. Soc. Chim. Fr., (1966) 3150.
26a P.M. Henry, J. Amer. Chem. Soc., 86 (1964) 3246.

26b P.M. Henry, J. Amer. Chem. Soc., 94 (1972) 4437.
26c I.I. Moiseev, D.G. Levanda and M.N. Vargaftik, J. Amer. Chem. Soc., 96 (1974) 1003.
27a P.M. Henry, J. Org. Chem., 32 (1967) 2575.
27b P.M. Henry, J. Amer. Chem. Soc., 94 (1972) 7305.
28a A. Aguilo, Advan. Organometal. Chem., 5 (1967) 321.
28b C.W. Bird, Transition Metal Intermediates in Organic Synthesis, Logos Press/Academic Press, New York, 1967, Chap. 4.
29 B.R. James and G.L. Rempel, Can. J. Chem., 46 (1968) 571.
30 J.E. Byrd and J. Halpern, J. Amer. Chem. Soc., 95 (1973) 2586.
31 P. Abley, J.E. Byrd and J. Halpern, J. Amer. Chem. Soc., 95 (1973) 2591.
32 R.J. Ouellette, G. Kordosky, C. Levin and S. Williams, J. Org. Chem., 34 (1969) 4104.
33 A. South and R.J. Ouellette, J. Amer. Chem. Soc., 90 (1968) 7064.
34 R.M. Barter and J.S. Littler, J. Chem. Soc. B, (1967) 205.
35 D. Benson and L.H. Sutcliffe, Trans. Faraday Soc., 55 (1959) 2107.
36 M.H. Dean and G. Skirrow, Trans. Faraday Soc., 54 (1958) 849.
37 M.S. Kharasch, A. Fono and W. Nudenberg, J. Org. Chem., 15 (1950) 763.
38 V. Stannett and R.B. Mesrobian, J. Amer. Chem. Soc., 72 (1950) 4125.
39 J.T. Martin and R.G.W. Norrish, Proc. Roy. Soc. London Ser. A, 220 (1953) 322.
40 R. Criegee, L. Kraft and B. Rank, Ann. Chem., 507 (1933) 159.
41 R. Criegee, E. Hoger, G. Huber, P. Kruck, F. Marktscheffel and H. Schellenberger, Ann. Chem., 599 (1956) 81.
42 D. Benson and N. Fletcher, Talanta, 13 (1966) 1207.
43 A.S. Perlin, J. Amer. Chem. Soc., 76 (1954) 5505.
44 J. Halpern and S.M. Taylor, Discuss. Faraday Soc., 29 (1960) 174.
45 M. Levy and M. Szwarc, J. Amer. Chem. Soc., 76 (1954) 5981.
46 V. Franzen and R. Edens, Angew. Chem., 73 (1961) 579.
47 R.O.C. Norman and M. Poustie, J. Chem. Soc. B, (1968) 781.
48 M.S. Kharasch, H.N. Friedlander and W.H. Urry, J. Org. Chem., 16 (1951) 533.
49 W.A. Mosher and C.L. Kehr, J. Amer. Chem. Soc., 75 (1953) 3172.
50 N. Dhar, J. Chem. Soc. London, 111 (1917) 690.
51 V.P. Kazakov, A.I. Mateeva, A.M. Erenburg and B.I. Peshchevitskii, Russ. J. Inorg. Chem. Engl. Transl., 10 (1965) 563.
52 Y. Pocker and B.C. Davis, J. Amer. Chem. Soc., 95 (1973) 6216.

53a   J.S. Littler, J. Chem. Soc. London, (1962) 827.
53b   J.S. Littler, J. Chem. Soc. London, (1964) 2722.
54   J.M. Davidson and C. Triggs, Chem. Ind. London, (1966) 457.
55   R. van Helden and G. Verberg, Rec. Trav. Chim. Pays-Bas, 84
       (1965) 1263.
56   J.M. Davidson and C. Triggs, J. Chem. Soc. A, (1968) 1324.
57   Y. Fujiwara, I. Moritani, M. Matsuda and S. Terranishi, Tetrahedron
       Lett., (1968) 3863.
58   E.W. Stern, in G.N. Schrauzer (Ed.), Transition Metals in
       Homogeneous Catalysis, Dekker, New York, 1971, p. 109.

Chapter 4

OXIDATIONS BY CHROMIUM(VI) AND MANGANESE(VII)

1. GENERAL FEATURES OF THE OXIDANTS

(a) Chromium(VI)

Various forms of chromium(VI) are used in practice, but by far the most common is chromium trioxide. The compound is easily available and dissolves in water and in some organic solvents, e.g. acetic anhydride. Because of difficulties over the solubility of some organic substrates in water, mixed solvents are often employed as reaction media. Since Cr(VI) is such a powerful oxidant, restrictions arise in the selection of the co-solvent and the usual choice is acetic acid. Other compounds of Cr(VI), e.g. chromyl chloride and chromyl acetate, are occasionally encountered in kinetic studies. Chromyl chloride (a deep red liquid) is best prepared by reaction of dichromate with sodium chloride and sulphuric acid. Chromyl acetate is formed by the action of chromium trioxide on acetic anhydride.

The equilibria involved in aqueous solutions of Cr(VI) are [1a, 1b]

$$H_2CrO_4 \rightleftharpoons H^+ + HCrO_4^- \qquad K_1 = 1.21 \text{ mol } l^{-1}$$
$$HCrO_4^- \rightleftharpoons H^+ + CrO_4^{2-} \qquad K_2 = 3 \times 10^{-7} \text{ mol } l^{-1}$$

together with the dimerisation step

$$2 \, HCrO_4^- \rightleftharpoons Cr_2O_7^{2-} + H_2O \qquad K_d = 98 \text{ l mol}^{-1}$$

where the equilibrium constants are quoted for 25°C. The predominant species of Cr(VI) in 10^{-4} M to 10^{-2} M solutions of pH 2—4 are $HCrO_4^-$ and $Cr_2O_7^{2-}$; under these conditions, H_2CrO_4 and CrO_4^{2-} are present in negligible amounts. The development of stopped-flow techniques has enabled the kinetics of the dimerisation step to be determined. The method employed by Pladziewicz and Espenson [2] was to bring about an abrupt two-fold dilution of a Cr(VI) solution at constant acidity, the resulting displacement of equilibrium being accompanied by a decrease in light absorbance since the molar extinction coefficient of $HCrO_4^-$ is much less than half that of $Cr_2O_7^{2-}$. In perchloric acid solutions, at a constant ionic strength of 1.0 M and over the range pH 2—4, the expression found for the rate of disappearance of $Cr_2O_7^{2-}$ is

$$-\frac{d\left[Cr_2O_7^{2-}\right]}{dt} = k_1\left[Cr_2O_7^{2-}\right]\left[H^+\right] - k_2\left[HCrO_4^-\right]^2\left[H^+\right]$$

where k_1 and k_2 are 6.35×10^3 l mol^{-1} s^{-1} and 6.22×10^5 l^2 mol^{-2} s^{-1}, respectively, at 25°C. The form of the rate law implies that dissociation of $Cr_2O_7^{2-}$ proceeds by a nucleophilic displacement of $HCrO_4^-$ by H_2O, the catalytic proton being associated with the bridging oxygen atom.

It is well-known that the oxidising power of acid Cr(VI) is markedly dependent on the medium. Furthermore, the ultraviolet spectrum of Cr(VI) changes as the acid medium is varied (Fig. 1). Both these effects can be attributed [3] to the reactions of the type

$$HCrO_4^- + H^+ + HCl = HOCrO_2 - Cl + H_2O$$
$$HCrO_4^- + H^+ + H_3PO_4 = HOCrO_2 - OPO_3H_2 + H_2O$$
$$HCrO_4^- + H^+ + H_2SO_4 = HOCrO_2 - OSO_3H + H_2O$$

If present in fairly high concentrations, chromium(VI) can

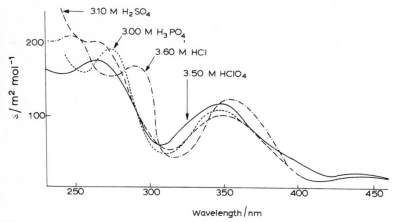

Fig. 1. Ultraviolet spectrum of chromium(VI) in various aqueous acids. (From Lee and Stewart [3], by courtesy of The American Chemical Society.)

be determined iodometrically; otherwise, it is convenient to use spectrophotometry (the full absorption spectrum is given in Fig. 8 on p. 191).

A few solid compounds of both Cr(V) and Cr(IV) have been prepared although they tend to decompose easily. In solution, most of the evidence for these intermediate oxidation states is kinetic in origin. However, a fairly long-lived species of Cr(V), green in colour and produced by the decomposition of Cr(VI) in strong potassium hydroxide solution, has been generated in sufficiently high concentrations to have been detected by electron spin resonance [4a, 4b]. More recently, substantial concentrations of Cr(V) have been trapped in 97% acetic acid and the visible and e.s.r. spectra recorded [5a, 5b]. Beattie and Haight [6] have provided a Frost diagram (Fig. 2) for the various chromium oxidation states. This consists of a

plot of potential against oxidation number, the slopes of the
tie lines giving the electrode potentials for the various couples.
It follows from the diagram that Cr(IV) will be reduced, with
ease, to Cr(III), will react with Cr(VI), and will have a strong
tendency to disproportionate.

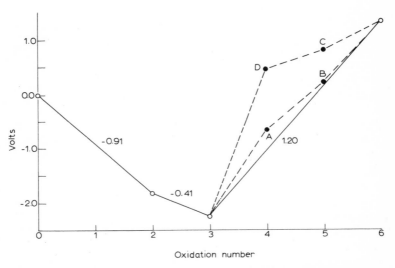

Fig. 2. Frost diagram for chromium. The points for Cr(IV) and Cr(V)
are estimated limits: point A from the capture of Cr(IV) by Ce(III),
point C from the reversible reduction of Cr(VI) to Cr(V) by Fe(II) and
the oxidation of iodide by Cr(VI), and points B and D from the capture
of Cr(IV) by Cr(VI). (From Beattie and Haight [6], by courtesy of
Interscience Publishers, John Wiley and Sons.)

(b) Manganese(VII)

Although permanganate is used in acidic, neutral, and
alkaline aqueous solutions, the reactive species is found to be

largely independent of the nature of the medium and is usually the simple oxoanion. However, protonation of the oxoanion can occur at high acidities

$$H^+ + MnO_4^- \rightleftharpoons HMnO_4$$

and the visual changes resulting from the addition of increasing amounts of acid to a permanganate solution (purple → red → brown → green) signify the gradual formation of the permanganic acid species; the process is half complete in 61% sulphuric acid [7] and 39% perchloric acid [8]. It is fortunate that exchange of oxygen between permanganate and water is slow in both neutral and basic solutions, as this allows mechanistic studies to be performed using oxygen-18 labelled reagent. However, oxygen exchange becomes rapid as the pH is reduced. Aqueous permanganate solutions are stable unless strongly acidic or strongly basic. The poor solubility of the reagent in a number of organic solvents (as reflected in the heterogeneous systems encountered in some organic syntheses) tends to restrict its usefulness although quite stable solutions can be prepared in acetic acid and moderately stable ones in dry acetone and *tert.*-butyl alcohol.

Permanganate can be determined by a variety of reagents, e.g. iodate, oxalate, oxalic acid, and iron(II), although its analysis is most frequently carried out by iodometry

$$2 MnO_4^- + 10 I^- + 16 H^+ = 2 Mn(II) + 5 I_2 + 8 H_2O$$
$$I_2 + 2 S_2O_3^{2-} = 2 I^- + S_4O_6^{2-}$$

where use is made of the overall reduction of Mn(VII) to the Mn(II) state. The very intense absorption of permanganate solutions in the visible region of the spectrum affords a very sensitive means of estimation. The spectrum is complex but the main (sharp) peaks occur at 526 and 546 nm (ϵ values,

154

240 and 238 m^2 mol^{-1}, respectively). In the ultraviolet, there
is a less intense and broader band at 311 nm ($\epsilon = 180$ m^2 mol^{-1}).
In contrast, Mn(VI) (as MnO_4^{2-}) and Mn(V) (as MnO_4^{3-}) are green
and blue, respectively, and have absorption peaks at 606 nm
and 667 nm.

In strongly basic solution, permanganate decomposes to
manganate and oxygen

$$4\ MnO_4^- + 4\ OH^- = 4\ MnO_4^{2-} + 2\ H_2O + O_2$$

and the initial step of the decomposition has been ascribed
[9] to

$$MnO_4^- + OH^- \rightleftharpoons MnO_4^{2-} + OH\cdot$$

Although doubt has been cast over the energetic feasibility of
such a step, it explains the observation that certain organic
compounds (e.g. *tert.*-butanol), unreactive in neutral or dilute
alkaline solution, are readily oxidised by strongly basic solutions of permanganate, oxidation proceeding by reactions of
the type

$$RH + OH\cdot \longrightarrow R\cdot + H_2O$$

without the production of oxygen [10].

The electron exchange of permanganate and manganate is
rapid and takes place by an outer-sphere process during which
the coordination shells of the two anions remain intact [11].
Alkali metal ions catalyse the exchange [12], in the order
$Cs^+ > K^+ > Na^+ > Li^+$, possibly via cation-bridged transition
states of the type $[O_3Mn-M-MnO_3^{2-}]^{\ddagger}$. Slow exchange
between permanganate and Mn(II) has been noted in perchloric
acid and nitric acid solutions. Adamson [13] has suggested
that exchange takes place essentially through the oxocation
species $Mn^{III}O^+$ and $Mn^{IV}O^{2+}$.

In general, manganate oxidations are slower than those of permanganate. In strongly basic solution, the rapid cleavage of 1,2-diols by manganate results in the formation of Mn(IV) (as MnO_2) whereas the oxidation of phenol gives rise to the blue-coloured Mn(V) species (hypomanganate, MnO_4^{3-}) [14]. Alcohols are oxidised slowly by manganate [15] as is the cyanide ion [16]. The oxidation of aromatic aldehydes is exceptional in that manganate and permanganate react with comparable rates [17]. A few reactions of manganate with inorganic reductants have been studied kinetically. The oxidation of periodate in alkaline solution [18]

$$2\ MnO_4^{2-} + H_3IO_6^{2-} = 2\ MnO_4^- + IO_3^- + 3\ OH^-$$

is second-order in manganate concentration and is believed to take place by

$$2\ MnO_4^{2-} \rightleftharpoons MnO_4^- + MnO_4^{3-}$$
$$MnO_4^{3-} + H_3IO_6^{2-} \longrightarrow MnO_4^- + IO_3^- + 3\ OH^-$$

although there is some indication that the first stage may be better expressed as

$$HMnO_4^- + MnO_4^{2-} \rightleftharpoons MnO_4^- + HMnO_4^{2-}$$

Hypochlorite is oxidised by manganate [19] according to

$$2\ MnO_4^{2-} + OCl^- + H_2O = 2\ MnO_4^- + Cl^- + 2\ OH^-$$

and the scheme suggested is

$$MnO_4^{2-} + H_2O \rightleftharpoons HMnO_4^- + OH^-$$
$$2\ HMnO_4^- \rightleftharpoons H_2MnO_4^- + MnO_4^-$$
$$H_2MnO_4^- + OCl^- \longrightarrow MnO_4^- + Cl^- + H_2O$$

Hypomanganate has a fair degree of stability in concentrated alkali but in moderately alkaline solutions disproportionation takes place as represented by

$$2\ MnO_4^{3-} + 2\ H_2O = MnO_4^{2-} + MnO_2 + 4\ OH^-$$

As far as can be gauged, the oxidising power of hypomanganate is less than that of manganate, the former oxidising primary and secondary alcohols slowly but having little reactivity towards tertiary alcohols, phenols, or alkenes [14]. It can be discerned as a transient in the oxidation of hydrogen peroxide by strongly alkaline permanganate, the colour of the reactant solution changing from purple to green to blue before brown MnO_2 is eventually deposited.

Manganese dioxide, the terminal product of the majority of permanganate oxidations, has uses as an oxidant in synthetic work in the form of a solid dispersion in acetone, ether, or chloroform. An example of its use is in the oxidation of α,β-unsaturated alcohols. The activity of the oxidant seems to depend upon its method of preparation.

2. REACTIONS WITH INORGANIC SPECIES

(a) Metal ions

There are a number of kinetic studies concerned with the oxidation of metal ions by Cr(VI) in perchlorate media. These include oxidation of V(IV) [20a—20c], V(III) [21], Fe(II) [22a, 22b], and Np(V) [23], viz.

$$HCrO_4^- + 3\ VO^{2+} + H^+ = Cr^{3+} + 3\ VO_2^+ + H_2O$$

$$HCrO_4^- + 3\ V^{3+} + H^+ = Cr^{3+} + 3\ VO^{2+} + H_2O$$

$$HCrO_4^- + 3\ Fe^{2+} + 7\ H^+ = Cr^{3+} + 3\ Fe^{3+} + 4\ H_2O$$

$$HCrO_4^- + 3\ NpO_2^+ + 7\ H^+ = Cr^{3+} + 3\ NpO_2^{2+} + 4\ H_2O$$

In all these, the basic mechanistic features can be represented as a sequence of three successive one-equivalent steps in which Cr(VI) is reduced to Cr(III) via the intermediates Cr(V) and Cr(IV).

$$\text{Cr(VI)} + \text{Red} \underset{k_{-1}}{\overset{k_1}{\rightleftharpoons}} \text{Cr(V)} + \text{Ox}$$

$$\text{Cr(V)} + \text{Red} \xrightarrow{k_2} \text{Cr(IV)} + \text{Ox}$$

$$\text{Cr(IV)} + \text{Red} \longrightarrow \text{Cr(III)} + \text{Ox} \qquad \text{rapid}$$

The full form of rate law for this mechanism is

$$-\frac{d\left[\text{Cr(VI)}\right]}{dt} = \frac{k_1 k_2 \left[\text{Cr(VI)}\right]\left[\text{Red}\right]^2}{k_2 \left[\text{Red}\right] + k_{-1}\left[\text{Ox}\right]}$$

which simplifies to

$$-\frac{d\left[\text{Cr(VI)}\right]}{dt} = k_1 \left[\text{Cr(VI)}\right]\left[\text{Red}\right]$$

when $k_2[\text{Red}] \gg k_{-1}[\text{Ox}]$, and to

$$-\frac{d\left[\text{Cr(VI)}\right]}{dt} = \frac{k_1 k_2 \left[\text{Cr(VI)}\right]\left[\text{Red}\right]^2}{k_{-1}\left[\text{Ox}\right]}$$

when $k_{-1}[\text{Ox}] \gg k_2[\text{Red}]$. The first limiting case applies to the oxidation of V(III) and the second to the reactions with V(IV) and Fe(II). In the case of the Np(V) reaction, the full form of rate expression is exhibited. In some of these reactions, e.g. with V(IV), the rates are independent of hydrogen ion concentration, implying initial steps of the type

$$\text{HCrO}_4^- + \text{VO}^{2+} + \text{H}_2\text{O} \rightleftharpoons \text{H}_3\text{CrO}_4 + \text{VO}_2^+ \qquad \text{rapid}$$

$$\text{H}_3\text{CrO}_4 + \text{VO}^{2+} \longrightarrow \text{Cr(IV)} + \text{VO}_2^+ \qquad \text{slow}$$

whereas in others a complex acid dependence is noted. In the case of the Fe(II) system, the observed third-order dependence on [H+] can be rationalised in terms of

$$HCrO_4^- + Fe^{2+} + 2\ H^+ \rightleftharpoons H_3CrO_4 + Fe^{3+} \qquad \text{rapid}$$

$$H_3CrO_4 + Fe^{2+} + H^+ \longrightarrow Cr(IV) + Fe^{3+} \qquad \text{slow}$$

Moreover, in the Cr(VI) + Fe(II) and Cr(VI) + V(IV) reactions a second term, second-order in Cr(VI), has been detected under certain conditions and this has been taken as indicating the participation of $Cr_2O_7^{2-}$.

Chromium(VI) reacts with Cr(II) in aqueous perchloric acid to give Cr^{3+} and a polymeric species (presumed to be $Cr_2(OH)_2^{4+}$) in the ratio [24a, 24b] of 2 to 1. When radio-chromium is used as a tracer, activity is found in both products. A compatible mechanism is

$$\overset{*}{Cr}(VI) + Cr(II) \longrightarrow \overset{*}{Cr}(V) + Cr^{3+}$$

$$\overset{*}{Cr}(V) + Cr(II) \begin{cases} \xrightarrow{\sim 80\%} \overset{*}{Cr}(IV) + Cr^{3+} \\ \xrightarrow[\sim 20\%]{} \overset{*}{Cr}{}^{3+} + Cr(IV) \end{cases}$$

$$\overset{*}{Cr}(IV) + Cr(II) \longrightarrow \overset{*}{Cr}(OH)_2 Cr^{4+}$$

$$Cr(IV) + Cr(II) \longrightarrow Cr(OH)_2 Cr^{4+}$$

It is interesting to note that the rate-controlling stage of the Ce(IV) + Cr(III) reaction (p. 21) has been identified as

$$Ce(IV) + Cr(IV) \longrightarrow Ce(III) + Cr(V)$$

and it may well be that the slowness of the Cr(V)—Cr(IV) interconversion stems from a necessity to change coordination number from 4 to 6. Further evidence is provided by the kinetics of exchange of Cr(VI) and Cr(III) in acid solution which suggest that the barrier to exchange is the slow conversion of Cr(V) to Cr(IV) [25].

Recently, Espenson and Wang [26] have studied the oxidation of U(IV) by Cr(VI) in perchlorate media

$$2\ HCrO_4^- + 3\ U^{4+} + 2\ H^+ = 2\ Cr^{3+} + 3\ UO_2^{2+} + 2\ H_2O$$

and have concluded that the reaction takes place without the active participation of U(V), proceeding instead by the sequence

$$Cr(VI) + U(IV) \longrightarrow Cr(IV) + U(VI) \qquad slow$$

$$Cr(IV) + Cr(VI) \longrightarrow 2\ Cr(V) \qquad rapid$$

$$Cr(V) + U(IV) \longrightarrow Cr(III) + U(VI) \qquad rapid$$

Permanganate oxidises ferrocyanide quantitatively to ferricyanide. An increase in rate is brought about by increasing acid concentration. Over the pH range 5—6 in solutions buffered with potassium phosphates, the slow stage is believed by Rawoof and Sutter [27] to involve an ion-pair formed with K^+

$$K^+ + Fe(CN)_6^{4-} \rightleftharpoons KFe(CN)_6^{3-} \qquad rapid$$

$$MnO_4^- + KFe(CN)_6^{3-} \longrightarrow MnO_4^{2-} + KFe(CN)_6^{2-} \qquad slow$$

At higher acidities (pH 1—2) protonated species take part

$$H^+ + Fe(CN)_6^{4-} \rightleftharpoons HFe(CN)_6^{3-} \qquad rapid$$

$$H^+ + HFe(CN)_6^{3-} \rightleftharpoons H_2Fe(CN)_6^{2-} \qquad rapid$$

$$MnO_4^- + H_2Fe(CN)_6^{2-} \longrightarrow MnO_4^{2-} + H_2Fe(CN)_6^- \qquad slow$$

The reaction between permanganate and tris(1,10-phenanthroline)iron(II), viz.

$$MnO_4^- + 5\ Fe(phen)_3^{2+} + 8\ H^+ = Mn(II) + 5\ Fe(phen)_3^{3+} + 4\ H_2O$$

is complex as indicated by the rate law

$$-\frac{(\tfrac{1}{5})\,d\left[Fe(phen)_3^{2+}\right]}{dt} = \frac{k_1\left[MnO_4^-\right]\left[Fe(phen)_3^{2+}\right]^2}{k_2\left[Fe(phen)_3^{3+}\right] + k_3\left[Fe(phen)_3^{2+}\right]}$$

for constant hydrogen ion concentration. Hicks and Sutter [28] interpret their kinetic data in terms of the following mechanism.

$$MnO_4^- + Fe(II) \rightleftharpoons MnO_4^{2-} + Fe(III)$$

$$HMnO_4 + Fe(II) \rightleftharpoons HMnO_4^- + Fe(III)$$

$$MnO_4^{2-} + Fe(II) \longrightarrow MnO_4^{3-} + Fe(III)$$

$$HMnO_4^- + Fe(II) \longrightarrow HMnO_4^{2-} + Fe(III)$$

(b) Non-metallic species

In the initial stages of the reaction between Cr(VI) and iodide ion in acid solution, the observed rate law is [29]

$$-\frac{d\left[I^-\right]}{dt} = k\left[HCrO_4^-\right]\left[I^-\right]^2\left[H^+\right]^2$$

in agreement with the sequence

$$Cr(VI) + I^- \rightleftharpoons Cr(VI),I^-\ \text{complex}$$

$$Complex + I^- \rightleftharpoons Cr(IV) + I_2$$

$$Cr(IV) + Cr(VI) \rightleftharpoons 2\ Cr(V)$$

$$Cr(V) + I^- \longrightarrow Cr(III) + IO^-$$

$$2\ H^+ + IO^- + I^- = H_2O + I_2$$

It is postulated that the hydrogen ion dependence develops as a result of the participation of the species $H_3CrO_4^+$. As the reaction progresses, the rate decreases and this may be due to the formation of molecular iodine (which is involved in a back reaction with Cr(IV)). Permanganate in acid solution oxidises both iodide and bromide ions

$$MnO_4^- + 5 X^- + 8 H^+ = Mn(II) + \tfrac{5}{2} X_2 + 4 H_2O$$

In the case of iodide, a two-part rate law applies [30]

$$-d\left[MnO_4^-\right]\big/dt = k_1\left[MnO_4^-\right]\left[I^-\right] + k_2\left[MnO_4^-\right]\left[I^-\right]\left[H^+\right]$$

The first term is attributed to the process

$$MnO_4^- + I^- \rightleftharpoons O_3MnOI^{2-} \qquad \text{rapid}$$
$$O_3MnOI^{2-} + HOH \longrightarrow HOI + HMnO_4^{2-} \qquad \text{slow}$$

and the second acid-dependent term to the analogous sequence

$$MnO_4^- + I^- \rightleftharpoons O_3MnOI^{2-} \qquad \text{rapid}$$
$$O_3MnOI^{2-} + H_3O^+ \longrightarrow HOI + H_2MnO_4^- \qquad \text{slow}$$

Thus H_2O and H_3O^+ are seen to compete for the intermediate. Hypoiodous acid is reduced to (complexed) iodine as represented by the net reaction

$$HOI + H_3O^+ + 2 I^- = I_3^- + 2 H_2O \qquad \text{rapid}$$

and the Mn(V) species ($HMnO_4^{2-}$ or $H_2MnO_4^-$) oxidises further iodide or disproportionates. Oxidation of bromide by permanganate has an overall resemblance to the oxidation of iodide with the important difference that the first stage involves an extra proton

$$MnO_4^- + H^+ + Br^- \rightleftharpoons HO_3MnOBr^- \qquad \text{slow}$$

$$HO_3MnOBr^- + H_3O^+ \longrightarrow HOBr + H_3MnO_4 \qquad \text{rapid}$$

$$HO_3MnOBr^- + H_2O \longrightarrow HOBr + H_2MnO_4^- \qquad \text{rapid}$$

Lawani and Sutter [31] have suggested that the involvement or non-involvement of the proton is decided by the potential of the substrate couple relative to that of the MnO_4^-/MnO_4^{2-} couple. If, as in the case of I_2/I^- and $Fe(CN)_6^{3-}/Fe(CN)_6^{4-}$, the substrate couple has a potential less than that for MnO_4^-/MnO_4^{2-} (+0.564 V) then the reaction proceeds without the need of a proton. However, if the potential of the substrate couple is greater than +0.564 V, as occurs with Br_2/Br^-, $Fe(phen)_3^{3+}/Fe(phen)_3^{2+}$, and Ce(IV)/Ce(III), then the reaction is thermodynamically unfeasible without the assistance of a proton in the initial step.

Under conditions of excess reductant in the pH range 4.2—5.0, Cr(VI) oxidises sulphite to a mixture of sulphate and dithionate as represented by the net reaction

$$2\,HCrO_4^- + 4\,HSO_3^- + 6\,H^+ = 2\,Cr^{3+} + 2\,SO_4^{2-} + S_2O_6^{2-} + 6\,H_2O$$

As a mechanism, Haight et al. [32] advocate the sequence

$$HCrO_4^- + HSO_3^- \rightleftharpoons CrSO_6^{2-} + H_2O$$

$$HSO_3^- + H^+ \rightleftharpoons SO_2 + H_2O$$

$$SO_2 + CrSO_6^{2-} \longrightarrow O_2SOCrOSO_2^{2-} \qquad \text{slow}$$

$$CrO_2(SO_3)_2^{2-} + 4\,H_2O + 2\,H^+ = SO_4Cr^{III}\,(H_2O)_5^+ + SO_3^-$$

$$2\,SO_3^- \longrightarrow S_2O_6^{2-}$$

which is in accord with the rate expression

$$-\frac{d\left[Cr(VI)\right]}{dt} = \frac{k\left[Cr(VI)\right]\left[S(IV)\right]^2\left[H^+\right]}{1 + k'\left[S(IV)\right]}$$

When Cr(VI) is in excess over sulphite, all S(IV) is oxidised to S(VI)

$$2\ HCrO_4^- + 3\ HSO_3^- + 5\ H^+ = 2\ Cr^{3+} + 3\ SO_4^{2-} + 5\ H_2O$$

and the following reactions are brought into play.

$$SO_3^- + HCrO_4^- \longrightarrow SO_4^{2-} + Cr(V)$$

$$H^+ + HCrO_4^- + CrSO_6^{2-} \longrightarrow O_3CrOCrOSO_2^{2-} + H_2O$$

$$O_3CrOCrOSO_2^{2-} \longrightarrow 2\ Cr(V) + SO_4^{2-}$$

$$Cr(V) + S(IV) \longrightarrow Cr(III) + S(VI)$$

Thiosulphate is oxidised by Cr(VI) to sulphate and tetra-thionate with variable stoichiometry [33]. At pH 4, the intermediate $CrS_2O_6^{2-}$, formed initially by the equilibrium [34]

$$HCrO_4^- + H^+ + S_2O_3^{2-} \rightleftharpoons CrS_2O_6^{2-} + H_2O$$

is destroyed by reaction with further thiosulphate ion as is indicated by the rate law

$$-\frac{d\left[Cr(VI)\right]}{dt} = k\left[Cr(VI)\right]\left[S_2O_3^{2-}\right]^2\left[H^+\right]$$

At high concentrations of thiosulphate relative to Cr(VI) (conditions under which tetrathionate is the favoured product), the slow stage is thought to be a two-equivalent reaction of the type [35]

$$CrS_2O_6^{2-} + HS_2O_3^- \longrightarrow Cr(IV) + S_4O_6^{2-}$$

which is followed by the rapid steps

$$Cr(IV) + S_2O_3^{2-} \longrightarrow Cr(III) + S_2O_3^-$$

$$2 S_2O_3^- \longrightarrow S_4O_6^{2-}$$

In strongly basic solution, permanganate oxidises cyanide quantitatively to cyanate

$$2 MnO_4^- + CN^- + 2 OH^- = 2 MnO_4^{2-} + CNO^- + H_2O$$

with a simple second-order rate law given by [16]

$$-\frac{d[MnO_4^-]}{dt} = k[MnO_4^-][CN^-]$$

Since tracer experiments with ^{18}O-labelled permanganate reveal that transfer of oxygen occurs from the oxidant to the cyanate product, it is likely that the initial step is

$$MnO_4^- + CN^- \rightleftharpoons [O_3Mn\text{--}O\text{--}CN^{2-}]^{\ddagger} \longrightarrow MnO_3^- + OCN^-$$

which is followed by a series of rapid steps summarised by

$$MnO_3^- + MnO_4^- + 2 OH^- = 2 MnO_4^{2-} + H_2O$$

In less basic solution (pH $<$ 12), cyanogen and carbon dioxide are formed in addition to cyanate and the rate law becomes more complex, viz.

$$-\frac{d\left[MnO_4^-\right]}{dt} = k\left[MnO_4^-\right]\left[CN^-\right]^2\left[H^+\right]$$

$$= k'\left[MnO_4^-\right]\left[HCN\right]\left[CN^-\right]$$

and little oxygen-18 is co-opted into the cyanate product. In this situation, a hydride transfer mechanism is considered appropriate

$$HCN + CN^- \rightleftharpoons H(CN)_2^-$$

$$MnO_4^- + H(CN)_2^- \longrightarrow HMnO_4^{2-} + (CN)_2$$

Carbon dioxide results from the oxidation of cyanogen and cyanate is formed in varying amounts by hydrolysis.

$$(CN)_2 + 2\,OH^- = CN^- + CNO^- + H_2O$$

Oxidation of molecular hydrogen by Cr(VI) can be brought about only in the presence of a catalyst. With Cu(II), the initial stages, represented [36a, 36b] by

$$Cu^{2+} + H_2 \rightleftharpoons CuH^+ + H^+$$

$$CuH^+ + Cu^{2+} \rightleftharpoons 2\,Cu^+ + H^+$$

are followed by the rapid regeneration of Cu^{2+} according to the net equation

$$Cr(VI) + 3\,Cu^+ = Cr(III) + 3\,Cu^{2+}$$

With Ag(I) as a catalyst, the initial steps are [37]

$$Ag^+ + H_2 \rightleftharpoons AgH + H^+$$

$$2\,Ag^+ + H_2 \longrightarrow 2\,AgH^+$$

Permanganate oxidises molecular hydrogen in acid and alkaline solutions, respectively, according to [38]

$$2\,MnO_4^- + 3\,H_2 + 2\,H^+ = 2\,MnO_2 + 4\,H_2O$$

$$2\,MnO_4^- + H_2 + 2\,OH^- = 2\,MnO_4^{2-} + 2\,H_2O$$

The rate law is second-order and a simple two-equivalent process seems likely, either

$$MnO_4^- + H_2 \longrightarrow HMnO_4^{2-} + H^+$$

or

$$MnO_4^- + H_2 \longrightarrow MnO_4^{3-} + 2\,H^+$$

The system is subject to catalysis by silver(I) when the rate-determining step is thought to be

$$MnO_4^- + H_2 + Ag^+ \longrightarrow MnO_4H^- + AgH^+$$

or

$$AgMnO_4 + H_2 \longrightarrow MnO_4^{2-} + AgH^+ + H^+$$

Reduction of permanganate by carbon monoxide takes place in both acid

$$2\,MnO_4^- + 3\,CO + H_2O = 2\,MnO_2 + 3\,CO_2 + 2\,OH^-$$

and alkaline solution

$$2\,MnO_4^- + CO + 4\,OH^- = 2\,MnO_4^{2-} + CO_3^{2-} + 2\,H_2O$$

with a simple second-order rate law [39], implying the slow stage to be

$$MnO_4^- + CO \longrightarrow O_3MnO - CO^-$$

This is followed by hydrolysis of the intermediate

$$O_3MnO - CO^- + H_2O \longrightarrow MnO_4^{3-} + CO_2 + 2 H^+$$

The reduction is catalysed by Ag^+ and Hg^{2+} ions. In the latter case, it appears that an intermediate, with the structure $[-Hg-CO-OMnO_3]$, takes part in the reaction.

Hydrogen peroxide reacts with Cr(VI) in acid solution to produce a blue peroxo complex, $CrO_5.H_2O$, whose rate of formation at moderate acidities is given by [40]

$$rate = k \left[HCrO_4^-\right] \left[H_2O_2\right] \left[H^+\right]$$

The favoured mechanism has, as a rate-determining step, a reaction between H_2CrO_4 and peroxide

$$HCrO_4^- + H^+ \rightleftharpoons H_2CrO_4$$

$$H_2CrO_4 + H_2O_2 \longrightarrow H_2CrO_5 + H_2O \qquad \text{slow}$$

$$H_2CrO_5 + H_2O_2 \longrightarrow CrO_5 \cdot H_2O + H_2O \qquad \text{rapid}$$

The alternative sequence, involving a protonated peroxide species

$$H_2O_2 + H^+ \rightleftharpoons H_3O_2^+$$

$$H_3O_2^+ + HCrO_4^- \longrightarrow H_2CrO_5 + H_2O \qquad \text{slow}$$

$$H_2CrO_5 + H_2O_2 \longrightarrow CrO_5 \cdot H_2O + H_2O \qquad \text{rapid}$$

is deemed less likely. No stable peroxy complex is formed by permanganate. Instead, peroxide is oxidised readily to oxygen and water. In acid solution, all the oxygen liberated is derived from the peroxide [41]. Furthermore, oxidation of D_2O_2 in D_2O solution is reported [42] to be considerably slower than oxidation of H_2O_2 in H_2O. Thus the oxygen—oxygen bond would seem to remain intact.

Cr(VI) oxidises hydrazine to nitrogen [43], hydroxylamine to nitrite and/or nitrous oxide (depending on the conditions) [44], and nitrous acid to nitrate [45]. All these systems have been studied kinetically.

A number of oxidations by Cr(VI) are inherently slow unless induced by the addition of small amounts of a third species [46]. Probably the best-known example is the reaction between Cr(VI) and iodide ions

$$2\ HCrO_4^- + 6\ I^- + 14\ H^+ = 2\ Cr^{3+} + 3\ I_2 + 8\ H_2O$$

This, in dilute acid solution and at low concentrations, is extremely slow but addition of Fe(II) brings about a marked acceleration

$$HCrO_4^- + Fe(II) + 2\ I^- + 7\ H^+ = Cr^{3+} + Fe(III) + I_2 + 4H_2O$$

Under the same conditions, the reaction

$$HCrO_4^- + 3\ Fe(II) + 7\ H^+ = Cr^{3+} + 3\ Fe(III) + 4\ H_2O$$

is rapid but that between Fe(III) and iodide

$$2\ Fe(III) + 2\ I^- = 2\ Fe(II) + I_2$$

is very slow. Obviously, Fe(II) cannot be described as a catalyst in the reaction since it is consumed in the primary stage

$$Cr(VI) + Fe(II) \; \rightleftharpoons \; Cr(V) + Fe(III)$$

and is not regenerated. The role of Fe(II) as inductor can be understood on the basis of the steps

$$Cr(V) + I^- \longrightarrow Cr(III) + IO^-$$
$$IO^- + I^- + 2H^+ = I_2 + H_2O$$

Induced reactions of this type are often described in terms of an induction factor which is defined [47] as the ratio of the number of equivalents of the acceptor (iodide) oxidised to the number of equivalents of inductor (Fe(II)) oxidised. In the case of the $Cr(VI) + Fe(II) + I^-$ system, the induction factor approaches a limiting value of 2 as the concentration of iodide ions is increased, i.e. as Fe(II) competes less and less favourably with iodide for the Cr(V) intermediate

$$Cr(V) + Fe(II) \longrightarrow Cr(IV) + Fe(III)$$
$$Cr(IV) + Fe(II) \longrightarrow Cr(III) + Fe(III)$$

In a number of oxidations of organic substrates, the induced oxidation of iodide has been used as a diagnostic test for the presence of a Cr(V) intermediate. Another familiar example is the reaction between Cr(VI) and Mn(II), induced by arsenite

$$2HCrO_4^- + 2H_3AsO_3 + Mn(II) + 6H^+ =$$
$$2Cr^{3+} + 2H_3AsO_4 + MnO_2 + 4H_2O$$

which has been discussed in terms of the sequence

$$Cr(VI) + As(III) \; \rightleftharpoons \; Cr(IV) + As(V)$$
$$Cr(IV) + Mn(II) \longrightarrow Cr(III) + Mn(III)$$
$$2\,Mn(III) \longrightarrow Mn(II) + Mn(IV)$$

Here, one equivalent of Mn(II) is oxidised for every two equivalents of As(III) and the limiting value of the induction factor is therefore 0.5.

3. OXIDATION OF HYDROCARBONS

Oxidation of aryl alkanes by chromic acid in glacial acetic acid has been investigated kinetically by Slack and Waters [48a, 48b] and Ogata et al. [49]. Diphenylmethane gives benzophenone as the major product, along with minor amounts of tetraphenylethane and traces of benzoic acid, phenol, and succinic acid; triphenylmethane yields triphenylcarbinol together with a lesser amount of benzophenone; and toluene is oxidised to benzoic acid. In 95% aqueous acetic acid and in the presence of an acid catalyst, the order of reactivity found by Wiberg and Evans [50] is triphenylmethane > diphenylmethane > ethylbenzene > toluene > methylcyclohexane > cyclohexane. The oxidation of diphenylmethane exhibits a pronounced kinetic isotope effect when deuterium is substituted for hydrogen, indicating that cleavage of the carbon— hydrogen bond is the slow stage. It seems reasonable to assume that the radical $R_3C\cdot$ is produced along with a Cr(V) intermediate.

Oxidation of aryl alkanes to aldehydes by chromyl chloride in an inert solvent like carbon tetrachloride is known as Étard's reaction. Kinetically, the reaction rate is first-order in each reactant [51a—51c]. Triphenylmethane and diphenylmethane are oxidised more rapidly than toluene, the order of reactivity being approximately 1000 : 100 : 1. Kinetic isotope effects indicate C—H bond breaking as the rate-determining step [52] and one possible mechanism is the chain

$$R_3CH + CrO_2Cl_2 \longrightarrow R_3C\cdot + HCrO_2Cl_2$$

$$R_3C\cdot + CrO_2Cl_2 \longrightarrow R_3C-O-CrOCl_2$$

$$R_3C-O-CrOCl_2 + R_3CH \longrightarrow R_3C-O-Cr(OH)Cl_2 + R_3C\cdot$$

Chlorinated hydrocarbons are formed as by-products. During the course of the reaction, a complex is produced which, under certain conditions, appears as a brown precipitate and, in the case of toluene, has the empirical formula $C_6H_5CH_3.2CrO_2Cl_2$. On heating, it loses hydrogen chloride and is converted into $C_6H_5CH_2.Cr_2O_4Cl_3$; on treatment with water, benzaldehyde is formed in good yield. There has been much speculation over the structure of the Étard complex. Electron spin resonance data suggest that both Cr atoms are in a +4 oxidation state [53]. This idea receives backing from the evidence of magnetic susceptibility measurements. In all probability, the complex has the symmetrical structure

$$C_6H_5-\underset{\underset{OCr(OH)Cl_2}{|}}{\overset{\overset{OCr(OH)Cl_2}{|}}{C}}-H$$

Loss of hydrogen chloride would then result in a ring structure containing a Cr—O—Cr bond

The original formulation [54] of the structure of the Étard complex in which one atom is Cr(IV) and the other Cr(VI), is now considered unlikely.

Toluene and ring-substituted toluenes are oxidised by chromium trioxide in acetic acid; the rates (determined iodometrically) at 70°C are first-order in substrate and second-order in Cr(VI) [49]. In terms of relative reactivity, the sequence followed is toluene $> p$-Br $> p$-Cl $> p$-CN $> p$-NO$_2$.

Information on the oxidation of alkanes by permanganate in aqueous solution is restricted because of solubility difficulties. Most studies have been concerned with the oxidation of various saturated chain compounds containing an inert carboxylic acid group, the presence of which imparts the necessary solubility. According to Wiberg and Fox [55], the mechanism of oxidation (of, for example, 4-methylhexanoic acid) is analogous to that for the chromic acid oxidation of hydrocarbons, involving initial abstraction of a hydrogen atom. In acetic acid, substituted toluenes are oxidised to the appropriate benzoic acid and ethylbenzene to acetophenone [56]. These systems do not permit a rigid kinetic analysis since the nature of the oxidant species remains unidentified.

Conversion of tertiary hydrogen to hydroxyl by oxidation with permanganate in concentrated basic solution is an important synthetic reaction. Typical is the oxidation of γ-phenylvaleric acid

$$MnO_4^- + Ar\overset{\overset{\displaystyle H}{|}}{\underset{\underset{\displaystyle CH_3}{|}}{C}}CH_2CH_2CO_2^- \xrightarrow{\ OH^-\ } MnO_4^{2-} + Ar\overset{\overset{\displaystyle OH}{|}}{\underset{\underset{\displaystyle CH_3}{|}}{C}}CH_2CH_2CO_2^-$$

which obeys a simple second-order rate law and shows a large kinetic isotope effect when the hydrogen at the γ position is replaced by deuterium [57]. A hydrogen atom abstraction by the oxidant is indicated as the rate-determining step. Two possibilities are open to the protonated permanganate ion and free radical so produced. The pair might recombine within the solvent cage to produce an ester species containing a Mn—O bond which could then undergo cleavage. Alternatively, on diffusing apart, the radical could be oxidised to a carbonium ion which would then react directly with hydroxide ion. In either case, the net effect is the production of a tertiary hydroxy compound. The reaction sequence can be generalised as follows.

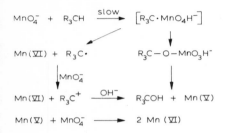

Olefins are oxidised by chromic acid with a rate which is first-order in both olefin and Cr(VI) concentrations [58]. Some olefins are oxidised more quickly in the presence of oxygen. For acetic acid media, the oxidation products are a mixture of diols, monoacetates, and diacetates as well as epoxides. The oxidation rate is little affected by steric factors in contrast to the oxidation of olefins by palladium(II) chloride where steric effects are paramount. On the other hand, the rate of oxidation by Cr(VI) is increased by increasing the number of alkyl substituents, whilst remaining insensitive to their position on the double bond, e.g. *cis*- and *trans*-2-butenes and isobutene react with comparable rates. In this context, there is a close resemblance to chlorine and bromine addition to olefins, and to epoxidation, but a marked difference to acid-catalysed hydration, hydroxymercuration [59], or thallium(III) oxidation [60] (since in these reactions isobutene is $\sim 10^3$ to 10^4 times more reactive than the 2-butenes). In the latter cases, the reactions are usually held to proceed through a carbonium ion intermediate whereas in the Cr(VI)— olefin system, the rate-determining stage is viewed as a symmetrical attack of the oxidant on the double bond leading to epoxide formation. The transition state in acetic acid media may be imagined as

where Cr(VI) is present as the monoacetate form. In aqueous media, the comparable transition states would be

Direct evidence has been obtained for the presence of an epoxide (cyclohexene oxide) in the Cr(VI) oxidation of cyclohexene in glacial acetic acid [61].

Alkenes react generally with permanganate as is evidenced by the well-known Baeyer test for unsaturation. Permanganate oxidation of unsaturated carboxylic acids has been studied by Wiberg and Geer [62]. In basic solution (above pH 12), permanganate is reduced to manganate and the major organic product is the diol. At a lower pH, ketol formation and cleavage occurs and Mn(IV) is formed. Kinetically, the initial rate of oxidation is first-order in both substrate and oxidant concentrations but is independent of acidity. It follows that the various products must derive from a common intermediate whose subsequent reactions are influenced by pH. Transfer of oxygen-18 from labelled alkaline permanganate to the oleate ion suggests that a cyclic ester is formed [63]. The relative rates of reaction for a series of olefins demonstrate a general lack of sensitivity to electronic effects. In a later publication, Wiberg et al. [64] have presented direct spectrophotometric evidence for two manganese(V) intermediates in the permanganate oxidation of crotonic acid (Fig. 3), the kinetic data being consistent with the scheme

$$MnO_4^- + CH_3CH=CHCOO^- \xrightarrow{k_1} A$$
$$A + OH^- \xrightarrow{k_2} B$$

where k_1 and k_2 are, respectively, 110 and 10 l mol^{-1} s^{-1} at 0°C.

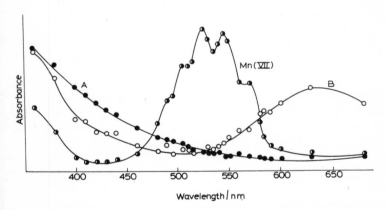

Fig. 3. Absorption spectra of permanganate and the two manganese(V) species (A and B) produced in the oxidation of crotonic acid as determined by an analysis of kinetic data. Absorbance values for B are multiplied by a factor of two. (From Wiberg et al. [64], by courtesy of The American Chemical Society.)

Looked at in more detail, the reaction is envisaged as proceeding according to

where A and B are converted to manganese(VI) by permanganate.

Chromyl chloride in carbon tetrachloride solution rapidly oxidises styrenes with a simple rate law given by [65]

$$-\frac{d[CrO_2Cl_2]}{dt} = k[CrO_2Cl_2][styrene]$$

Stopped-flow techniques show the formation of a clearly-defined 1:1 adduct. A rate-determining attack of the oxidant on the C=C bond is indicated. These results are substantiated by a study of the oxidation of a series of 15 alkenes, all of which are shown to form chromyl chloride—alkene adducts [66]. A cyclic three-membered ring transition state is likely, viz.

In general, acetylenic compounds are oxidised very rapidly by permanganate. A typical case is the oxidation of acetylene-dicarboxylic acid which takes place cleanly in acid solution according to the equation

$$2\ MnO_4^- + HOOCC \equiv CCOOH + 6H^+ = 2\ Mn^{2+} + 4\ CO_2 + 4\ H_2O$$

In the course of reduction from Mn(VII) to Mn(II), detectable amounts of Mn(III) are formed as shown by stopped-flow traces taken at 250 nm [67a, 67b]. The reaction can be held at this stage of oxidation by the addition of pyrophosphate which acts as a scavenger, complexing strongly with Mn(III) and thereby reducing its reactivity. In acid solution, with the dicarboxylic acid in excess over oxidant, oxalic acid is produced as an intermediate and oxygen-18 experiments reveal that complete transfer of oxygen atoms occurs from oxidant to substrate. Formation of a Mn(V) intermediate of cyclic structure has been suggested as the initial stage of the process, the rate being first-order with respect to permanganate and substrate concentrations.

$$II \longrightarrow 2\ Mn(IV) + 2\ (CO_2H)_2$$

$$4\ Mn(IV) + I \longrightarrow 5\ Mn(III) + (CO_2H)_2 + 2\ CO_2$$

$$2\ Mn(III) + (CO_2H)_2 \longrightarrow 2\ Mn(II) + 2\ CO_2 + 2\ H^+$$

4. OXIDATION OF ALCOHOLS AND GLYCOLS

Probably the most common application of chromium(VI) in organic synthesis is in the oxidation of alcohols to carbonyl compounds. Primary alcohols are oxidised in the first place to aldehydes but these may be oxidised further to the corresponding carboxylic acid. However, a side reaction between the aldehyde and unreacted alcohol can give rise to the formation of an ester via the hemiacetal, viz.

$$RCH_2OH + RCHO \rightleftharpoons R-\underset{\underset{OH}{|}}{\overset{\overset{H}{|}}{C}}-OCH_2R \xrightarrow{Cr(VI)} R-\underset{\underset{O}{\|}}{C}-OCH_2R$$

In the case of secondary alcohols, the straight-forward oxidation to the appropriate ketone can be complicated by cleavage occurring, for example

$$C_6H_5-\underset{\underset{H}{|}}{\overset{\overset{OH}{|}}{C}}-C(CH_3)_3 \xrightarrow{Cr(VI)} C_6H_5-\underset{\underset{O}{\|}}{C}-C(CH_3)_3 + C_6H_5CHO + (CH_3)_3COH$$

The oxidation of primary and secondary alcohols by chromic acid is described [68] by the general rate law

$$\text{Rate} = k_1' \left[HCrO_4^- \right] \left[alcohol \right] \left[H^+ \right] + k_2' \left[HCrO_4^- \right] \left[alcohol \right] \left[H^+ \right]^2$$

Kinetic isotope effects indicate that the rate-controlling stage in the reaction is the breaking of the carbon—hydrogen bond in the alcohol [69a, 69b]. In the case of isopropanol, Watanabe and Westheimer [70] proposed in 1949 that the gross features of the reaction could be represented as

$$Cr(VI) + R_2CHOH \longrightarrow R_2C=O + Cr(IV)$$

$$Cr(IV) + Cr(VI) \longrightarrow 2 Cr(V)$$

$$Cr(V) + R_2CHOH \longrightarrow R_2C=O + Cr(III)$$

In this scheme, ketone can be formed in two different ways, both of which are two-equivalent processes. Furthermore, Cr(VI) brings about only one-third of the total oxidation, the major part resulting from the attack of the Cr(V) intermediate on the substrate. Naturally, the first step is rate-determining and, because of ample evidence for the formation of acid chromate esters, can be detailed as

$$R_2CHOH + HCrO_4^- + H^+ \rightleftharpoons R_2CHOCrO_3H + H_2O$$

$$R_2CHOCrO_3H + H^+ \rightleftharpoons R_2CHOCrO_3H_2^+$$

$$R_2CHOCrO_3H \xrightarrow{k_1} R_2C=O + Cr(IV)$$

$$R_2CHOCrO_3H_2^+ \xrightarrow{k_2} R_2C=O + Cr(IV)$$

This is seen to be compatible with the observed rate law.

Oxidation of tertiary alcohols can be effected by Cr(VI). The oxidation rates are independent of the concentration of oxidant, and Roček [71a, 71b] has put forward the suggestion that the initial step is the acid-catalysed dehydration of the alcohol to give the respective alkene which then suffers oxidation.

As noted above, the chromic acid oxidation of secondary alcohols can result in cleavage as well as ketone formation. Various schemes have been proposed. In 1956, Westheimer and co-workers [72] demonstrated that it is Cr(V) or Cr(IV), rather than Cr(VI), which is responsible for the cleavage reaction and amended their general mechanism to take account of these alternatives, represented by

$$Cr(VI) + RCHOHR' \longrightarrow Cr(IV) + ketone$$

$$Cr(IV) + Cr(VI) \longrightarrow 2\,Cr(V)$$

$$2\,Cr(V) + 2\,RCHOHR' \longrightarrow 2\,Cr(III) + 2\,RCH=O + 2\,R'OH$$

or

$$Cr(VI) + RCHOHR' \longrightarrow Cr(IV) + ketone$$

$$Cr(IV) + RCHOHR' \longrightarrow RCH=O + R'\cdot + Cr(III)$$

$$Cr(IV) + R'\cdot + H_2O \longrightarrow Cr(V) + R'OH + H^+$$

$$Cr(V) + RCHOHR' \longrightarrow Cr(III) + ketone$$

Of these two mechanisms, the first appears more acceptable in that it is compatible with the observation that approximately two-thirds of the products result from cleavage, whereas the second scheme would give rise to only 33% cleavage. It should be noted, however, that the second stage of the first mechanism is hardly viable since the equilibrium position of the reaction is far to the left ($K = 4 \times 10^{-14}$, as estimated from electrochemical data [73]). A reappraisal of the situation has been provided by Roček and Radkowsky [74a, 74b] and they have established that oxidative cleavage of cyclobutanol is brought about by Cr(IV) and not by Cr(V). The method was based on the previous findings of Espenson [20b] for the reaction between Cr(VI) and V(IV) for which the accepted mechanism is

$$Cr(VI) + V(IV) \rightleftharpoons Cr(V) + V(V) \qquad rapid$$

$$Cr(V) + V(IV) \longrightarrow Cr(IV) + V(V) \qquad slow$$

$$Cr(IV) + V(IV) \longrightarrow Cr(III) + V(V) \qquad rapid$$

When cyclobutanol is added to the reacting V(IV) + Cr(VI) system, only the cleavage product (4-hydroxybutyraldehyde) is formed. Furthermore, the rate of disappearance of Cr(VI) is not affected by addition of alcohol. Thus it follows that Cr(IV) reacts with the alcohol and not Cr(V) since the latter is involved with Cr(VI) in a rapid pre-equilibrium.

Accordingly, the evidence is that cyclobutanol is oxidised as follows.

$$Cr(VI) + V(IV) \rightleftharpoons Cr(V) + V(V) \qquad \text{rapid}$$
$$Cr(V) + V(IV) \longrightarrow Cr(IV) + V(V) \qquad \text{slow}$$
$$Cr(IV) + C_4H_7OH \longrightarrow Cr(III) + R\cdot$$
$$R\cdot + V(V) \longrightarrow HOCH_2CH_2CH_2CHO + V(IV)$$

As a consequence of their results on the Cr(VI) + V(IV) system, Roček and Radkowsky have proposed that the Cr(VI) oxidation of secondary alcohols proceeds by

$$Cr(VI) + RCHOHR' \longrightarrow Cr(IV) + \text{ketone} \qquad \text{slow}$$
$$2\,Cr(IV) + 2\,RCHOHR' \longrightarrow 2\,Cr(III) + 2\,RCH{=}O + 2\,R'\cdot$$
$$2\,Cr(VI) + 2\,R'\cdot \longrightarrow 2\,Cr(V) + 2\,R'OH$$
$$2\,Cr(V) \longrightarrow Cr(VI) + Cr(IV)$$

This, an adaption of the second Westheimer scheme, accounts for the observed 67% cleavage whilst placing Cr(V) secondary in importance as an intermediate. Confirmatory evidence [75] for the decisive role played by Cr(IV) is that the oxidative cleavage of a group of 2-aryl-1-phenylethanols by Cr(VI) (in acetic acid) is a one-equivalent process (as is their oxidation by Ce(IV) [76]). Again, approximately two-thirds cleavage is realised (independent of varying Cr(III) concentration) and trapping experiments with oxygen indicate that cleavage results in a benzyl radical. Furthermore, acrylamide and acrylonitrile are polymerised when added to a reacting Cr(VI) + isopropanol system [73]. Since Cr(IV) and Cr(V) have been shown not to lead to polymer formation, the presence of an intermediate free radical is indicated, generated by the reaction of Cr(IV) with the substrate. Rahman and Roček [77] have extended the work on cyclobutanol by carrying out a general survey of Cr(IV) reactivity towards primary and secondary alcohols. Oxidation of 2-deutero-2-propanol displays an isotope effect commensurate with that found for other one-equivalent oxidations (involving Ce(IV), Co(III), and Mn(III)), a result which suggests that C—H bond cleavage is rate-determining. *Tert.*-butyl alcohol is unreactive towards Cr(IV).

Chromium(V) has been detected and characterised in the Cr(VI) + isopropanol system by means of its visible and electron spin resonance spectra [5b]. Ester formation is indicated by a decrease of absorbance at 385 nm with time and the formation and decay of the Cr(V) intermediate by absorbance measurements at 510 nm paralleled by the change in intensity of the e.s.r. signal (at $g = 1.9805$). The rate of formation of acetone is in accord with Cr(IV), not Cr(V), functioning as the active oxidant (Fig. 4) [78].

The presence of either cerium(III) or cerium(IV), even at concentrations as low as 10^{-7} M, reduces the rate of oxidation of isopropanol by Cr(VI). Furthermore, little or no cleavage of phenyl *tert.*-butyl carbinol occurs when these ions are present. Roček and co-workers [79] explain these results by the mechanism

$$\begin{aligned}
\text{Cr(VI)} + \text{RCHOHR}' &\longrightarrow \text{Cr(IV)} + \text{ketone} \qquad \text{slow} \\
\text{Cr(IV)} + \text{Ce(III)} &\longrightarrow \text{Cr(III)} + \text{Ce(IV)} \\
\text{Cr(IV)} + \text{Ce(IV)} &\longrightarrow \text{Cr(V)} + \text{Ce(III)} \\
\text{Cr(V)} + \text{Ce(IV)} &\longrightarrow \text{Cr(VI)} + \text{Ce(III)}
\end{aligned}$$

in which the Cr(IV) intermediate is removed either by reduction with Ce(III) or by oxidation with Ce(IV). In other words, Ce(III) catalyses the disproportionation of Cr(IV)

$$3\,\text{Cr(IV)} \xrightarrow{\text{Ce(III)}} 2\,\text{Cr(III)} + \text{Cr(VI)}$$

and the net reaction has the stoichiometry

$$2\,\text{Cr(VI)} + 3\,\text{RCHOHR}' = 2\,\text{Cr(III)} + 3\,\text{(ketone)}$$

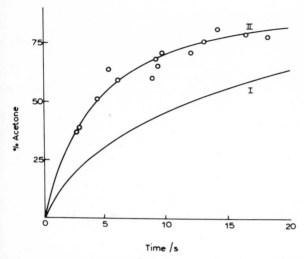

Fig. 4. Rate of formation of acetone in the chromic acid oxidation of isopropanol in 97% acetic acid. Curve I represents the calculated values for a scheme whose initial steps are

$$Cr(VI) + R_2CHOH \rightarrow Cr(IV) + R_2C=O$$
$$Cr(VI) + Cr(IV) \rightarrow 2\,Cr(V)$$

Curve II assumes the initial steps to be

$$Cr(VI) + R_2CHOH \rightarrow Cr(IV) + R_2C=O$$
$$Cr(IV) + R_2CHOH \rightarrow Cr(III) + R_2\dot{C}OH$$

$[Cr(VI)] = 0.005\,M$, $[R_2CHOH] = 0.126\,M$, $[H^+] = 0.0125\,M$, ionic strength = 0.184 M, 15°C. (From Wiberg and Mukherjee [78], by courtesy of The American Chemical Society.)

Details of the interaction between Cr(VI) (as chromic acid dissolved in 97% acetic acid) and isopropanol have been provided by Wiberg and co-workers [5a, 78] in terms of equilibrium and rate constants (at 15°C) for the sequence

$$AcOCrO_2OH \rightleftharpoons AcOCrO_2O^- + H^+ \qquad K_1 = 0.24 \text{ mol } l^{-1}$$
$$AcOCrO_2OH + ROH \rightleftharpoons ROCrO_2OH + AcOH \qquad K_2 = 115 \text{ l mol}^{-1}$$
$$ROCrO_2OH \rightleftharpoons ROCrO_2O^- + H^+ \qquad K_3 = 0.019 \text{ mol } l^{-1}$$
$$ROCrO_2OH + ROH \rightleftharpoons ROCrO_2OR + H_2O \qquad K_4 = 9.21 \text{ l mol}^{-1}$$

The equilibrium constant for dimer formation

$$2 HOCrO_2OAc \rightleftharpoons AcOCrO_2OCrO_2OH + HOAc$$

is estimated as ~85 l mol^{-1} for the same conditions. The changes in the relative proportions of monoester, monoester anion, and diester at varying alcohol concentrations are depicted in Fig. 5.

Hasan and Roček [80] report an interesting study of the chromic acid oxidation of a *mixture* of oxalic acid and isopropanol. When present alone, each substrate is oxidised at only a moderate rate but when present together both are oxidised very rapidly. The full rate expression (determined spectrophotometrically in perchloric acid solution) comprises three terms

$$\text{rate} = k_1 [HCrO_4^-][H^+]^2[ROH] + k_2 [HCrO_4^-][(COOH)_2][ROH] + k_3 [HCrO_4^-][(COOH)_2]^2$$

Of these, the second term predominates and represents the co-oxidation reaction. In this, one molecule of oxalic acid *and* one molecule of alcohol are oxidised by one Cr(VI) species in a single three-equivalent reaction to give one molecule each of acetone and carbon dioxide together with a $\cdot CO_2^-$ (or $\cdot CO_2H$) radical. In the present of acrylonitrile, which acts as a radical scavenger, acetone and CO_2 are formed in equal amounts, independent of the relative concentrations of reactants, thus indicating that isopropanol and oxalic acid undergo a two-

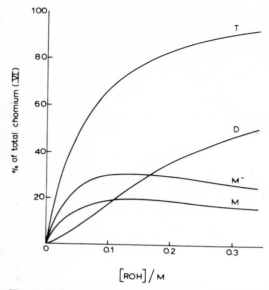

Fig. 5. Relative concentrations of monoester (M), monoester anion (M⁻), diester (D), and total esters (T) as a function of isopropanol concentration. (From Wiberg and Mukherjee [78], by courtesy of The American Chemical Society.)

equivalent oxidation and a one-equivalent oxidation, respectively. Since chromic acid and oxalic acid form a complex which is probably cyclic in type, Hasan and Roček suggest that an ester-type intermediate is formed, which then undergoes a rate-determining decomposition.

A deuterium isotope effect confirms that the C—H bond in the alcohol is broken in the rate-determining stage. In the absence of a radical trap, the $\cdot CO_2^-$ radical might then dimerise or be oxidised further by Cr(VI) yielding CO_2 and Cr(V), which in turn could then be reduced by either of the substrates [81], i.e.

$$Cr(VI) + \cdot CO_2^- \longrightarrow CO_2 + Cr(V)$$
$$Cr(V) + R_2CHOH \longrightarrow Cr(III) + R_2CO + H_2O$$
$$Cr(V) + (COOH)_2 \longrightarrow Cr(III) + 2\,CO_2 + H_2O$$

If all the Cr(V) reacted with isopropanol, the stoichiometry would correspond to

$$2\,Cr(VI) + 2\,R_2CHOH + (COOH)_2 = 2\,Cr(III) + 2\,R_2CO + 2\,CO_2 + 6\,H^+$$

whereas if all Cr(V) was consumed by oxalic acid, the stoichiometry would be

$$2\,Cr(VI) + R_2CHOH + 2\,(COOH)_2 = 2\,Cr(III) + R_2CO + 4\,CO_2 + 6\,H^+$$

Thus the ratio of CO_2 to acetone would be expected to vary between limits of 4:1 and 1:1, depending on the relative con-

187

centrations of oxalic acid and alcohol. This prediction has been verified experimentally (Fig. 6). In more detail, the relative reactivity of Cr(V) towards the two substrates can be assessed. If m is the mole fraction of Cr(V) reacting with isopropanol then Cr(V) is removed according to

$$m\ Cr(V) + m\ R_2CHOH \xrightarrow{k_1} m\ Cr(III) + m\ R_2CO$$

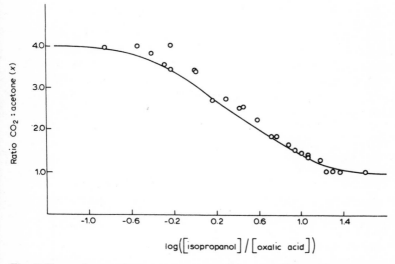

Fig. 6. Dependence of the ratio of oxidation products (x) on the substrate ratio in the chromic acid co-oxidation of isopropanol and oxalic acid. The curve is calculated from the equation

$$x = \frac{(k_1/k_2)([R_2CHOH]/[(COOH)_2]) + 2}{(k_1/k_2)([R_2CHOH]/[(COOH)_2]) + 0.5}$$

using $k_1/k_2 = 0.27$. (From Hasan and Roček [81], by courtesy of The American Chemical Society.)

and

$$(1-m)\,Cr(\underline{V}) + (1-m)\,(COOH)_2 \xrightarrow{k_2} (1-m)\,Cr(\underline{III}) + 2(1-m)\,CO_2$$

where

$$m = \frac{k_1\left[R_2CHOH\right]}{k_1\left[R_2CHOH\right] + k_2\left[(COOH)_2\right]}$$

For each two Cr(VI) species reduced, the total yield of acetone is $1 + m$ and the total yield of CO_2 is $2 + 2(1-m) = 2(2-m)$. Defining the observed ratio of products as x, i.e.

$$x = \frac{CO_2\ \text{yield}}{\text{acetone yield}} = \frac{2(2-m)}{1+m}$$

it follows that

$$\frac{4-x}{2(x-1)} = \frac{k_1\left[R_2CHOH\right]}{k_2\left[(COOH)_2\right]}$$

Figure 7 shows the linear plot obtained of the left-hand side of this equation versus the ratio of reactants. The slope of the plot can be equated to the ratio of the rate constants k_1/k_2 (= 0.27 at 25°C), the latter value reflecting the relative reactivities of the two substrates (at 25°C oxalic acid is ~3.7 times more reactive towards Cr(V) than is isopropanol).

Recent work has shown that the Cr(V) intermediate, being relatively long-lived, is amenable to investigation by electron spin resonance techniques. The e.s.r. spectrum, consisting of two narrow and intense lines which grow and then decay at the same rate, is unlike that for Cr(III) (a single broad line) and there is no possibility of confusion with Cr(IV) since the latter gives an e.s.r. spectrum only at low temperatures.

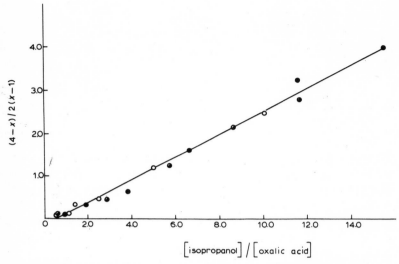

Fig. 7. Relative rates of chromium(V) oxidation of isopropanol and oxalic acid at perchloric acid concentrations of ○, 0.063 M; ●, 0.125 M; and ◉, 0.63 M. (From Hasan and Roček [81], by courtesy of The American Chemical Society.)

Srinivasan and Roček [82] hold that two Cr(V) species are produced and that the overall reaction between Cr(V) and oxalic acid can be viewed as taking place through a fairly stable monoxalato complex of Cr(V) which breaks down to give Cr(III) only after the addition of a second substrate molecule.

The u.v. and visible spectrum of the Cr(V) species has been constructed indirectly and is given in Fig. 8 along with spectra of Cr(VI) and Cr(III) for comparison. No ring cleavage occurs during the co-oxidation by Cr(VI) of cyclobutanol and oxalic acid [83]. Chromium(IV) therefore seems to be absent, the cyclobutanone product arising from reaction between Cr(V) and the alcohol.

Although tetraalkoxides of tertiary alcohols have been known for some time, a stable Cr(IV) alkoxide of a secondary alcohol has been prepared only recently [84]. The complex, tetrakis(3,3-dimethyl-2-butoxy)chromium(IV), prepared from tetra-*tert.*-butoxychromium(IV) and 3,3-dimethyl-2-butanol, is surprisingly stable and resists decomposition for as long as 12 h at 85°C in dioxan solvent.

Both primary and secondary alcohols are oxidised by permanganate under basic or acidic conditions. Tertiary alcohols are immune to attack unless drastic conditions are applied, when degradation results. Primary alcohols give rise to aldehydes which in some cases may be oxidised further to the carboxylic acids. Secondary alcohols give ketones but, if these are enolisable, cleavage may occur with alkaline permanganate. Permanganate oxidation of alcohols is subject to catalysis by strong acids and strong bases. The effect of acids is attributed to protonation of the oxidant whereas formation

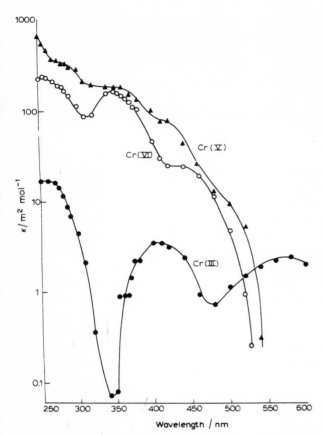

Fig. 8. Absorption spectra of chromium(VI), chromium(III), and chromium(V) in aqueous solution. (From Srinivasan and Roček [82], by courtesy of The American Chemical Society.)

of the alkoxide ion is held to be responsible for the activation brought about by strong bases. Two possibilities are open as regards the mechanism, both of which involve an interaction between the MnO_4^- and the alkoxide species in the form of a hydrogen abstraction [15,85]. In the first, a hydride ion is transferred

$$R_2CHO^- + MnO_4^- \longrightarrow R_2C{=}O + HMnO_4^{2-}$$

and in the second, a hydrogen atom

$$R_2CHO^- + MnO_4^- \longrightarrow R_2\dot{C}{-}O^- + HMnO_4^-$$

A substantial kinetic isotope effect for the oxidation of R_2CDOH signifies cleavage of a C—H bond. Of the two steps, the first is considered the less likely but evidence for its rejection is inconclusive. Oxidation of benzyl alcohol and its monosubstituted derivatives by acid permanganate has been examined in the presence of fluoride ions [86]. Without the latter, the reactions are autocatalytic. Transfer of a hydride ion from the alcohol carbon is proposed as the rate-determining stage.

Oxidation of ethanediol by chromic acid appears to be a normal alcohol-type reaction. However, increasing C-methylation increases the possibility of carbon—carbon bond cleavage [87] and, in the case of pinacol, cleavage (to give acetone) can amount to 70%. For all glycols studied, a simple second-order rate law is followed and strong retardation by added Mn(II) is noted except in the case of pinacol [88]. The latter is oxidised much more rapidly than ethanediol although the associated activation energy is larger than that for the simpler glycol. The consensus of opinion is that reaction proceeds by a mechanism similar to that generally accepted for glycol-cleavage by lead(IV) (p. 134), viz.

$$H^+ + HCrO_4^- + \begin{matrix} R_2C-OH \\ | \\ R_2C-OH \end{matrix} \overset{rapid}{\rightleftharpoons} \begin{matrix} R_2C-O \\ | \quad\quad Cr \\ R_2C-O \end{matrix} \overset{O}{\underset{O}{<}} + 2\,H_2O$$

slow

$$\begin{matrix} R_2C=O \\ + \\ R_2C=O \end{matrix} \quad + \quad Cr(IV)$$

As for Pb(IV) cleavage, the case for the intervention of a cyclic ester-type intermediate rests with rate differences for *cis* and *trans* isomers. In aqueous perchloric acid (at 30°C), *cis*-1,2-dimethyl-1,2-cyclopentanediol is oxidised to 2,6-heptane-dione some 17,000 times faster than its *trans* counterpart. At room temperature in 90% acetic acid, the *cis* isomer reacts 300 times faster than the *trans* isomer [89]. Little is known about the cleavage of glycols by permanganate. Reaction is slow compared with the oxidation of alkenes.

5. OXIDATION OF ALDEHYDES AND KETONES

In aqueous media, chromic acid oxidises both benzaldehyde [90a, 90b] and aliphatic aldehydes [91] with features closely resembling those for the oxidation of isopropanol. Accordingly, the mechanism generally accepted is

$$RCHO + H^+ + HCrO_4^- \rightleftharpoons R-\underset{H}{\overset{OH}{\underset{|}{\overset{|}{C}}}}-OCrO_3H \longrightarrow R-\overset{OH}{\overset{|}{C}}=O + Cr(IV)$$

with C—H bond cleavage as the slow stage. Using an approach similar to that employed for the oxidation of alcohols, a two-step process involving Cr(VI) and Cr(V) has been identified in the oxidation of benzaldehyde in 96% acetic acid [92]. The

Cr(V) intermediate has been characterised by means of its light absorption and e.s.r. signal. Aliphatic aldehydes are more readily oxidised by Cr(V) than are aromatic aldehydes.

Addition of benzaldehyde to a solution of chromium trioxide in acetic anhydride brings about a colour change from orange to dark brown. Analysis of the brown solution reveals that it still has oxidising power and that Cr(VI) has been reduced, not to Cr(III), but to Cr(IV). The organic product is benzoic acid which was determined by isotope dilution analysis. In the presence of excess aldehyde, the reaction sequence can be visualised in terms of two consecutive first-order processes [93]. In the first, Cr(V) is formed as an intermediate and in the second, Cr(V) is reduced to Cr(IV), i.e.

$$A \xrightarrow{k_1} B$$
$$B \xrightarrow{k_2} C$$

The concentrations of each species are then given by

$$[A] = [A]_0 \exp(-k_1 t)$$
$$[B] = \frac{[A]_0 k_1 [\exp(-k_2 t) - \exp(-k_1 t)]}{(k_2 - k_1)}$$

and

$$[C] = [A]_0 - [A] - [B]$$

Absorption spectra of B and C (Cr(V) and Cr(IV)), reconstructed from a computer analysis of the kinetic data, are given in Fig. 9 along with the spectra of chromyl acetate and chromium(III) acetate. Reacting mixtures of oxidant and substrate give an electron spin resonance spectrum and this has been attributed to the presence of Cr(V). Oxygen-18 experiments demonstrate that oxygen is transferred from the oxidant

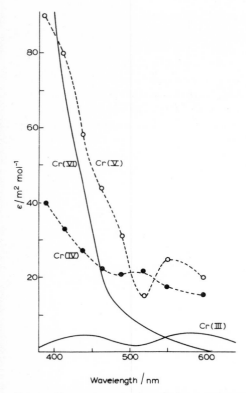

Fig. 9. Absorption spectra of chromyl acetate, chromium(III) acetate, chromium(V), and chromium(IV). (From Wiberg and Lepse [93], by courtesy of The American Chemical Society.)

to the aldehyde and this, coupled with the observation of a kinetic isotope effect, indicates

$$RCHO + (AcO)_2CrO_2 \longrightarrow R\dot{C}{=}O + (AcO)_2CrO_2H$$

as the first stage, followed by reaction of the two intermediates within the solvent cage to yield eventually benzoic acid, possibly through a mixed anhydride of the type $(AcO)_2\overset{O}{\underset{}{C}}r-O\overset{O}{\underset{}{C}}R$. However, the reaction is complicated by the possibility of interactions between Cr(V) and the aldehyde and between Cr(VI) and the radical, and also by complex formation between Cr(VI) and aldehyde. Addition of sodium acetate brings about a reduction in rate, stemming from a change in the reactive Cr(VI) species.

$$O_2Cr(OAc)_2 + OAc^- \longrightarrow AcOCrO_3^- + Ac_2O$$

Chromium(IV), generated by reaction of Cr(VI) with V(IV), has a high reactivity towards aldehydes [94]. Propionaldehyde is oxidised quantitatively to propionic acid and the observed deuterium isotope effect indicates a rate-determining transfer of a hydrogen atom from the aldehyde to the oxidant.

Pivaldehyde on oxidation gives a mixture of pivalic acid, *tert.*-butyl alcohol, and isobutylene, possibly via oxidation or decarbonylation of a radical intermediate.

The rate of oxidation of aromatic aldehydes by permanganate is greatly influenced by changes in pH. In buffered solution, the stoichiometry corresponds to

$$2\,MnO_4^- + 3\,RCHO = 2\,MnO_2 + 2\,RCOO^- + RCOOH + H_2O$$

In neutral solution, the oxidation is acid-catalysed, kinetically second-order, and takes place with the transfer of oxygen from the oxidant to the substrate. Furthermore, a comparison with the rate of oxidation of ArCDO reveals a significant kinetic isotope effect. Wiberg and Stewart [17] propose a mechanism in which an ester-type intermediate is formed with the aldehyde, the rate-determining stage being the loss of a proton from the ester. In basic solution, there is little deuterium isotope effect and most of the oxygen introduced into the aldehyde originates in the solvent. It seems clear that the mechanism is completely different from that which operates in acidic or neutral solution and a free radical chain may well be involved.

A stopped-flow study of the permanganate oxidation of the heterocyclic aldehyde furfural reveals two second-order pathways [95]. The first is independent of OH^- concentration and corresponds to the formation and subsequent decomposition of a permanganate ester, i.e.

$$R-\overset{\overset{O}{\|}}{C}-H + MnO_4^- \rightleftharpoons R-\overset{\overset{O^-}{\|}}{\underset{\underset{O-MnO_3}{|}}{C}}-H \xrightarrow{slow} R-\overset{\overset{O}{\|}}{C}-O^- + MnO_3^- + H^+$$

$$3\,MnO_3^- + H_2O = 2\,MnO_2 + MnO_4^- + 2\,OH^- \quad rapid$$

where $R = $. The second route is more important, is

dependent on OH⁻ concentration, and consists of the formation of a hydrate anion of the aldehyde which then transfers a hydride ion to the oxidant as shown by

$$R-\overset{\overset{O}{\|}}{C}-H + OH^- \rightleftharpoons R-\overset{\overset{O^-}{|}}{\underset{\underset{OH}{|}}{C}}-H$$

$$R-\overset{\overset{O^-}{|}}{\underset{\underset{OH}{|}}{C}}-H + MnO_4^- \xrightarrow{slow} R-\overset{\overset{O}{\|}}{C}-OH + HMnO_4^{2-} \xrightarrow{rapid} R-\overset{\overset{O}{\|}}{C}-O^- + MnO_3^- + H_2O$$

$$Mn(\text{V}) + Mn(\text{VII}) \longrightarrow 2\,Mn(\text{VI}) \quad rapid$$

Oxygen-18 experiments demonstrate that the solvent is the principal source of oxygen introduced into the substrate.

Observations on the oxidation of cyclohexanone by chromic acid suggest that enolisation precedes oxidation [96]. In perchloric acid media under conditions where oxidation of intermediate products is insignificant, the rate of the reaction is first-order in oxidant, substrate, and hydrogen ion concentrations (for solutions of constant ionic strength). Furthermore, the rate of oxidation is slower than the rate of enolisation of the ketone at the same acidity. A detailed consideration of deuterium isotope effects leads to the conclusion that the slow stage corresponds to the breakdown of the intermediate

The initial product of the overall oxidation is 2-hydroxycyclo-hexanone. Further oxidation results in 1,2-cyclohexanedione and cleavage gives rise to adipic acid [97]. Glutaric and succinic acids are formed also.

Kinetic proof for an enol intermediate has been obtained in the oxidation of isobutyrophenone and 2-chlorocyclo-hexanone by chromium trioxide [98]. Assuming the general mechanism

$$\text{ketone} \underset{k_K}{\overset{k_E}{\rightleftharpoons}} \text{enol} \xrightarrow[k]{CrO_3} \text{product}$$

to apply, it follows that

$$-\frac{d[Cr(VI)]}{dt} = \frac{kk_E[Cr(VI)][\text{ketone}]}{k_K + k[Cr(VI)]}$$

where k, k_E, and k_K are the rate constants for oxidation, enolisation and ketonisation, respectively. When the oxidation step is rate-controlling, then $k_K \gg k[Cr(VI)]$ and the rate is first-order in Cr(VI) concentration. On the other hand, when enolisation is rate-controlling, then $k[Cr(VI)] \gg k_K$ and the rate becomes independent of the concentration of Cr(VI). Under these latter conditions, the rate of disappearance of the oxidant will be the same as the rate of enolisation. The predicted change-over from a first-order to a zero-order dependence on Cr(VI) is depicted in Fig. 10.

The permanganate oxidation of acetone in basic solution is described by the third-order rate law [99]

$$-\frac{d[Mn(VII)]}{dt} = k[MnO_4^-][\text{acetone}][OH^-]$$

which is consistent with the mechanism

200

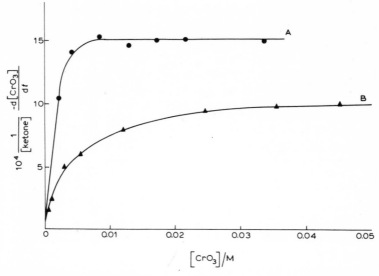

Fig. 10. Effect of initial concentration of chromic acid on the rate of oxidation at 30°C. A, isobutyrophenone in 0.50 M HClO₄ in 99% acetic acid and B, 2-chlorocyclohexanone in 1.0 M aqueous HClO₄. (From Roček and Riehl [98], by courtesy of The American Chemical Society.)

$$\text{acetone} + OH^- \rightleftharpoons \text{enolate} + H_2O$$
$$Mn(VII) + \text{enolate} \longrightarrow Mn(V) + \text{product} \qquad \text{slow}$$
$$Mn(VII) + Mn(V) \longrightarrow 2\,Mn(VI) \qquad \text{rapid}$$

Information on non-rate-determining steps has been obtained by product analysis. Solutions of oxidant and acetone, labelled with carbon-14, were allowed to react for a short time and then quenched with bisulphite. At this intermediate stage of the reaction, the products identified and estimated by isotope dilution analysis were acetol, lactic acid, pyruvic acid (formed via pyruvaldehyde), oxalic acid, and acetic acid. The rates of oxidation of some of these have been studied separately. The reaction sequence is presumed to be

$$CH_3\overset{O}{\overset{\|}{C}}CH_3 \longrightarrow CH_3\overset{O}{\overset{\|}{C}}CH_2OH \longrightarrow CH_3\overset{O}{\overset{\|}{C}}CHO \longrightarrow CH_3\overset{O}{\overset{\|}{C}}COO^- \longrightarrow CH_3\overset{O}{\overset{\|}{C}}O^-$$
$$\longrightarrow {}^-OOCCOO^-$$

where the intermediate formed by reaction of permanganate with enolate is most likely the CH_3COCH_2 radical, which in turn is oxidised further to pyruvaldehyde either directly or indirectly via acetol.

6. OXIDATION OF CARBOXYLIC ACIDS

At least three mechanisms are possible for the oxidation of formate ions by permanganate above pH 5. Firstly, the transfer of a hydride ion may occur in the rate-determining stage

$$MnO_4^- + HCOO^- \longrightarrow HMnO_4^{2-} + CO_2$$

Alternatively, the slow stage may consist of the transfer of an electron-pair with concurrent proton transfer to the solvent

$$MnO_4^- + HCOO^- + H_2O \longrightarrow MnO_4^{3-} + CO_2 + H_3O^+$$

A third possibility is that abstraction of a hydrogen atom takes place

$$MnO_4^- + HCOO^- \longrightarrow HMnO_4^- + \cdot CO_2^-$$

and this is followed by the destruction of the radical ion by the oxidant

$$MnO_4^- + \cdot CO_2^- \longrightarrow \left[O_3Mn-O-CO_2\right]^{2-} \begin{array}{c} \nearrow MnO_4^{2-} + CO_2 \\ \\ \searrow MnO_3 + CO_3^{2-} \end{array}$$

All these possibilities are in accord with the considerable primary hydrogen isotope effect observed for this reaction [100]. Discrimination on the grounds of the solvent isotope effect is not possible since the value found for $k(H_2O)/k(D_2O)$ is too close to unity [101]. However, the transfer of oxygen-18 from permanganate to the carbon dioxide product [100] is supporting evidence for the third mechanism. At higher acidities than pH 5, the reactive species is the formic acid molecule, not the formate ion, and the reaction is much slower. In strong sulphuric acid media, permanganic acid is formed and the rate increases again.

Formate bound to Co(III) is oxidised by permanganate [102]. The reduction of the latter is accompanied by reduction of Co(III) to Co(II)

$$MnO_4^- + 3 Co(NH_3)_5 \cdot OCHO^{2+} \longrightarrow 3 CO_2 + 3 Co^{2+} + MnO_2$$

but retention of the Co(III) centre is possible also as shown by

$$2 MnO_4^- + 3 Co(NH_3)_5 \cdot OCHO^{2+} \longrightarrow 3 CO_2 + 3 (NH_3)_5 Co \cdot OH_2^{3+} + 2 MnO_2$$

The rate expression

$$-\frac{d\left[Co(NH_3)_5 \cdot OCHO^{2+}\right]}{dt} = k \left[MnO_4^-\right]\left[Co(NH_3)_5 \cdot OCHO^{2+}\right]$$

applies and the ratio of unreduced to reduced cobalt(III) is proportional to the concentration of MnO_4^-. Replacement of the hydrogen in the formate group by deuterium results in a substantial reduction in rate and the initial stage is considered to be a hydrogen abstraction reaction

$$MnO_4^- + (NH_3)_5 Co^{III}(OCHO^-)^{2+} \longrightarrow HMnO_4^- + (NH_3)_5 Co^{III}(CO_2^-)^{2+}$$

which is followed by breakdown of the intermediate to give
either $Co^{2+} + CO_2$, or $(NH_3)_5Co^{III}.OH_2^{3+} + CO_2$. It is interesting
to note that one-equivalent oxidants (for example, Ce(IV))
bring about reduction of the Co(III) centre in oxalatopenta-
amminecobalt(III) whereas two-equivalent oxidants, like
chlorine, do not [103].

The Cr(VI)—oxalic acid system has been scrutinised by a
number of workers, notably, and most recently, by Hasan
and Roček [104]. The reaction (in perchlorate media) is first-
order in oxalic acid at low concentrations, changes to second-
order at moderate concentrations, and reverts back to first-
order at high concentrations. This behaviour is explicable in
terms of the formation of two Cr(VI)—oxalic acid complexes,
one a neutral cyclic anhydride and the other an open-chain
dianion. The latter 1:2 complex generates a monoanion intermediate which is considered then to break down to give
$Cr(H_2O)_6^{3+}$, three molecules of CO_2, and a $\cdot CO_2H$ radical in a
single three-equivalent step. The $\cdot CO_2H$ radicals can then
reduce Cr(VI) to give Cr(V), recombine to give oxalic acid, or
react to give formic acid and CO_2, viz.

$\cdot CO_2H + Cr(VI) \longrightarrow CO_2 + Cr(V)$

$2 \cdot CO_2H \longrightarrow (CO_2H)_2$ or $CO_2 + HCO_2H$

The familiar permanganate—oxalate reaction

$$2\,MnO_4^- + 5\,C_2O_4^{2-} + 16\,H^+ = 2\,Mn(II) + 10\,CO_2 + 8\,H_2O$$

has been the subject of extensive kinetic study [105]. It now seems quite clear that no direct reaction occurs between permanganate and oxalate. Added Mn(II) accelerates the reaction. Conversely, if Mn(II) is removed by complexing with fluoride, permanganate and oxalate have little or no tendency to react. In the absence of fluoride, Mn(II) is complexed with oxalate and the initial step in the reduction of permanganate is considered to be the one-equivalent process [106]

$$Mn^{VII}O_4^- + Mn^{II}C_2O_4 \longrightarrow Mn^{VI}O_4^{2-} + Mn^{III}C_2O_4^+$$

When small amounts of Mn(II) are present, manganate is reduced by reaction with oxalate

$$Mn(VI) + C_2O_4^{2-} \longrightarrow Mn(IV) + 2\,CO_2$$
$$2\,Mn(IV) + C_2O_4^{2-} \longrightarrow 2\,Mn(III) + 2\,CO_2$$

In the presence of appreciable amounts of Mn(II), manganate is reduced instead by reaction with Mn(II)

$$Mn(VI) + Mn(II) \longrightarrow 2\,Mn(IV)$$
$$Mn(IV) + Mn(II) \longrightarrow 2\,Mn(III)$$

When the concentration of oxalate is very low, the system becomes autocatalytic and this behaviour is ascribed to a reaction between Mn(III), complexed with oxalate, and permanganate

$$MnO_4^- + MnC_2O_4^+ \longrightarrow MnO_4^{2-} + Mn(IV)$$

In the normal course of events, Mn(III) is destroyed by reactions of the type

$$MnC_2O_4^+ \longrightarrow Mn(II) + CO_2 + \cdot CO_2^-$$
$$MnC_2O_4^+ + \cdot CO_2^- \longrightarrow Mn(II) + C_2O_4^{2-} + CO_2$$

It is thought that atmospheric oxygen may take part in the reaction through the formation of peroxo species

$$\cdot CO_2^- + O_2 \longrightarrow O_2CO_2^{-\cdot}$$
$$O_2CO_2^{-\cdot} + Mn(II) + 2H^+ \longrightarrow Mn(III) + H_2O_2 + CO_2$$

A kinetic study of the initial stages of the oxidation of lactic, malic, and mandelic acids by chromic acid in mixed aqueous perchloric acid—acetic acid media has shown that the rate is first-order in each of the reactant concentrations (Cr(VI) reacting as $HCrO_4^-$) and in hydrogen ion concentration [107]. Oxidation would seem to proceed by a mechanism similar to that formulated for the oxidation of secondary alcohols (p. 179).

This interpretation is in accord with the large primary kinetic isotope effect noted for α-deuteromandelic acid. 2-Hydroxy-methylpropanoic acid, which contains no α-hydrogen atom, is relatively inert to chromic acid.

Chromic acid oxidation of mandelic acid (to give benzaldehyde) is subject to catalysis by Mn(II) provided that the ratio [Mn(II)]/[Cr(VI)] is greater than 100:1; under these conditions the rate becomes independent of the concentration of Cr(VI) [108]. At lower concentrations than this, Mn(II) retards the reaction. Chromium(III) has no effect on the rate at high [Mn(II)]/[Cr(VI)] ratios even when added in hundredfold excess. No evidence was adduced for the presence of radicals in the reacting mixtures in that acrylamide neither affected the rate nor was it polymerised. Beckwith and Waters [108] advocate as a basis for their mechanism the scheme suggested by Roček and Radkowsky for the oxidation of cyclobutanol (p. 181). Chromium(VI) attacks the substrate to yield Cr(IV) which then reacts further. The substrate is also attacked by Mn(III), formed by

$$Cr(VI) + Mn(II) \rightleftharpoons Cr(V) + Mn(III)$$

At high Mn(II) concentrations, Cr(VI) is reduced by Mn(II) as well as by substrate; at low Mn(II) concentrations, the strong oxidant Cr(IV) is reduced to Cr(III) by Mn(II)

$$Cr(IV) + Mn(II) \rightleftharpoons Cr(III) + Mn(III)$$

Ethyl mandelate, a hydroxy-acid ester, is oxidised by chromic acid (in perchloric acid solution) with kinetics [109] closely resembling those for the oxidation of mandelic acid and secondary alcohols, predictably since all these substrates have a —CHOH grouping in common. A mechanism in which C—H bond rupture occurs, viz.

seems appropriate, the reaction rate being first-order in substrate, Cr(VI), and hydrogen ion concentrations with ethyl phenylglyoxylate as the major product.

Manganese(III) has been detected as a transient intermediate in the oxidation of maleic and fumaric acids by acid permanganate, stopped-flow traces at 250 nm showing the formation and decay of the species [110]. At this stage of reaction, the organic product is formyl(hydroxy)acetic acid. Further reduction of Mn(III) takes place with concurrent formation of hydroxymalonic, glyoxylic, and oxalic acids, the final products being Mn(II) and formic acid. A possible sequence of events is

$$2 \; Mn(III) + I \longrightarrow HOOC \cdot CHOH \cdot COOH + 2 \; Mn(II)$$

$$III \longrightarrow CHO \cdot COOH + CO_2 + Mn(III)$$
$$III \longrightarrow (COOH)_2 + HCOOH + Mn(III)$$

The rate of the first stage, a four-equivalent oxidation of the substrate, is first-order in both permanganate and substrate concentrations and is strongly dependent on pH [111]. The kinetics suggest that the rate-controlling step corresponds to the *cis* attack of oxidant on the double bond of the unsaturated acid giving rise to the formation of a cyclic intermediate of Mn(V).

This is short-lived since no direct evidence for its presence has been forthcoming from the application of stopped-flow techniques. The observation that fumaric acid is considerably more reactive than maleic acid is ascribed to steric factors operating in the latter case.

7. OXIDATION OF PHENOLS AND AMINES

Chromium(VI) oxidises quinol quantitatively according to

$$2\,HCrO_4^- + 3\,H_2Q + 8\,H^+ = 2\,Cr^{3+} + 3\,Q + 8\,H_2O$$

The reaction in perchloric acid—lithium perchlorate media has been followed in detail by a spectrophotometric method [112] and the rate law is

$$-\frac{d[Cr(VI)]}{dt} = k\left[Cr(VI)\right]\left[H_2Q\right]\left[H^+\right]^{1.25}$$

No evidence has been obtained for the presence of a chromium ester intermediate even with the use of stopped-flow techniques. The hydrogen ion dependence of the rate can be described alternatively by a power series of the type

$$k' = a + b\left[H^+\right] + c\left[H^+\right]^2$$

where the parameters b and c are the rate constants for

parallel reaction steps with activated complexes containing one and two hydrogen ions, respectively. The oxidation of 2,6-dinitrophenol by acid permanganate displays a pronounced induction period [113]. Behaviour similar to that encountered in the oxidation of oxalate (p. 204) has been noted in that Mn(II) increases the rate whereas fluoride ions have an inhibiting influence.

Aliphatic amines are oxidised by weakly basic permanganate according to the reactivity order tertiary > secondary > primary. In the case of trimethylamine, the final product is formaldehyde and the reaction seems best viewed as an electron transfer although hydrogen atom or hydride ion abstractions may well be involved in the oxidation of primary and secondary amines [114].

$$MnO_4^- + N(CH_3)_3 \rightleftharpoons H_3C-\overset{+}{\underset{\cdot}{N}}(CH_3)_2 + MnO_4^{2-} \qquad \text{slow}$$

$$H_3C-\overset{+}{\underset{\cdot}{N}}(CH_3)_2 \longrightarrow H^+ + H_2C=\!\!=\!\!N(CH_3)_2 \qquad \text{rapid}$$

$$H_2C=\!\!=\!\!N(CH_3)_2 \xrightarrow{MnO_4^-} H_2C=\!\!\overset{+}{N}(CH_3)_2 \qquad \text{rapid}$$

$$H_2C=\!\!\overset{+}{N}(CH_3)_2 + H_2O \longrightarrow H-\overset{\overset{\displaystyle O}{\|}}{C}-H + HN(CH_3)_2 + H^+ \qquad \text{rapid}$$

Benzylamine is oxidised by alkaline permanganate to give as final products benzaldehyde, benzoic acid, and benzamide although the first stage in the oxidation produces benzalimine [115].

$$2\,MnO_4^- + 3\,C_6H_5CH_2NH_2 = 2\,MnO_2 + 3\,C_6H_5CH=\!\!=\!\!NH + 2\,H_2O + 2\,OH^-$$

Reaction is minimal below pH 6, the rate increasing gradually over the pH range 6—8 and then increasing rapidly from pH 8 to pH 10 to reach a limiting value at pH 12. Kinetic data are consistent with a rate-determining reaction between oxidant and neutral amine, the oxidation of protonated amine being so slow as to be discounted. Products of the rate-determining

210

stage are either the radical $C_6H_5\overset{.}{C}HNH_2$ together with $HMnO_4^-$ (Mn(VI)), or the protonated imine $C_6H_5CH=NH_2^+$ and $HMnO_4^{2-}$ (Mn(V)). In strongly alkaline solution (> pH 12), it seems likely that a new step, involving OH^- ion, is introduced and the reaction is very rapid.

REFERENCES

1a J.Y. Tong and E.L. King, J. Amer. Chem. Soc., 75 (1953) 6180.
1b J.Y. Tong, Inorg. Chem., 3 (1964) 1804.
2   J.R. Pladziewicz and J.H. Espenson, Inorg. Chem., 10 (1971) 634.
3   D.G. Lee and R. Stewart, J. Amer. Chem. Soc., 86 (1964) 3051.
4a N. Bailey and M.C.R. Symons, J. Chem. Soc. London, (1957) 203.
4b A. Carrington, D.J.E. Ingram, D.S. Schonland and M.C.R. Symons, J. Chem. Soc. London, (1956) 4710.
5a K.B. Wiberg and H. Schäfer, J. Amer. Chem. Soc., 91 (1969) 927.
5b K.B. Wiberg and H. Schäfer, J. Amer. Chem. Soc., 91 (1969) 933.
6   J.K. Beattie and G.P. Haight, in J.O. Edwards (Ed.), Inorganic Reaction Mechanisms, Part 2, Interscience, 1972, p.97.
7   R. Stewart and M.M. Mocek, Can. J. Chem., 41 (1963) 1160.
8   N. Bailey, A. Carrington, K.A.K. Lott and M.C.R. Symons, J. Chem. Soc. London, (1960) 290.
9   M.C.R. Symons, J. Chem. Soc. London, (1953) 3956.
10 K.A.K. Lott and M.C.R. Symons, Discuss. Faraday Soc., 29 (1960) 205.
11 O.E. Meyers and J.C. Sheppard, J. Amer. Chem. Soc., 83 (1961) 4730.
12 L. Gjertsen and A.C. Wahl, J. Amer. Chem. Soc., 81 (1959) 1572.
13 A.W. Adamson, J. Phys. Colloid Chem., 55 (1951) 293.
14 J.S.F. Pode and W.A. Waters, J. Chem. Soc. London, (1956) 717.
15 R. Stewart, J. Amer. Chem. Soc., 79 (1957) 3057.
16 R. Stewart and R. van der Linden, Can. J. Chem., 38 (1960) 2237.
17 K.B. Wiberg and R. Stewart, J. Amer. Chem. Soc., 77 (1955) 1786.
18 M.W. Lister and Y. Yoshino, Can. J. Chem., 38 (1960) 2342.
19 M.W. Lister and Y. Yoshino, Can. J. Chem., 39 (1961) 96.
20a J.H. Espenson, J. Amer. Chem. Soc., 86 (1964) 1883.
20b J.H. Espenson, J. Amer. Chem. Soc., 86 (1964) 5101.
20c K.M. Davies and J.H. Espenson, J. Amer. Chem. Soc., 92 (1970) 1889.

21 K.M. Davies and J.H. Espenson, J. Amer. Chem. Soc., 92 (1970) 1884.
22a J.H. Espenson and E.L. King, J. Amer. Chem. Soc., 85 (1963) 3328.
22b J.H. Espenson, J. Amer. Chem. Soc., 92 (1970) 1880.
23 J.C. Sullivan, J. Amer. Chem. Soc., 87 (1965) 1495.
24a L.S. Hegedus and A. Haim, Inorg. Chem., 6 (1967) 664.
24b J.C. Kenny and D.W. Carlyle, Inorg. Chem., 12 (1973) 1952.
25 C. Altman and E.L. King, J. Amer. Chem. Soc., 83 (1961) 2825.
26 J.H. Espenson and R.T. Wang, Inorg. Chem., 11 (1972) 955.
27 M.A. Rawoof and J.R. Sutter, J. Phys. Chem., 71 (1967) 2767.
28 K.W. Hicks and J.R. Sutter, J. Phys. Chem., 75 (1971) 1107.
29 K.E. Howlett and S. Sarsfield, J. Chem. Soc. A, (1968) 683.
30 L.J. Kirschenbaum and J.R. Sutter, J. Phys. Chem., 70 (1966) 3863.
31 S.A. Lawani and J.R. Sutter, J. Phys. Chem., 77 (1973) 1547.
32 G.P. Haight, E. Perchonock, F. Emmenegger and G. Gordon, J. Amer. Chem. Soc., 87 (1965) 3835.
33 M.I. Edmonds, K.E. Howlett and B.L. Wedzicha, J. Chem. Soc. A, (1970) 2866.
34 I. Baldea and G. Niac, Inorg. Chem., 7 (1968) 1232.
35 I. Baldea and G. Niac, Inorg. Chem., 9 (1970) 110.
36a E. Peters and J. Halpern, J. Phys. Chem., 59 (1955) 793.
36b J. Halpern, E.R. MacGregor and E. Peters, J. Phys. Chem., 60 (1956) 1455.
37 A.H. Webster and J. Halpern, J. Phys. Chem., 61 (1957) 1239.
38 A.H. Webster and J. Halpern, Trans. Faraday Soc., 53 (1957) 51.
39 A.C. Harkness and J. Halpern, J. Amer. Chem. Soc., 83 (1961) 1258.
40 M. Orhanović and R.G. Wilkins, J. Amer. Chem. Soc., 89 (1967) 278.
41 A.E. Cahill and H. Taube, J. Amer. Chem. Soc., 74 (1952) 2312.
42 T.-L. Chang, Science, 100 (1944) 29.
43 G.P. Haight, T.J. Huang and B.Z. Shakashiri, J. Inorg. Nucl. Chem., 33 (1971) 2169.
44 N. Hlasivcova and I. Novak, Collect. Czech. Chem. Commun., 36 (1971) 2027.
45 D.A. Durham, L. Dozsa and M.T. Beck, J. Inorg. Nucl. Chem., 33 (1971) 2971.
46 L.J. Csányi, in C.H. Bamford and C.F.H. Tipper (Eds.), Comprehensive Chemical Kinetics, Vol. 7, Elsevier, Amsterdam, 1972, p. 510.
47 F.H. Westheimer, Chem. Rev., 45 (1949) 419.

48a   R. Slack and W.A. Waters, J. Chem. Soc. London, (1948) 1666.
48b   R. Slack and W.A. Waters, J. Chem. Soc. London, (1949) 599.
49   Y. Ogata, A. Fukui and S. Yuguchi, J. Amer. Chem. Soc., 74 (1952) 2707.
50   K.B. Wiberg and R.J. Evans, Tetrahedron, 8 (1960) 313.
51a   R.A. Stairs and J.W. Burns, Can. J. Chem., 39 (1961) 960.
51b   R.A. Stairs, Can. J. Chem., 40 (1962) 1656.
51c   R.A. Stairs, Can. J. Chem., 42 (1964) 550.
52   O.H. Wheeler, Can. J. Chem., 42 (1964) 706.
53   K.B. Wiberg and R. Eisenthal, Tetrahedron, 20 (1964) 1151.
54   I. Necsoiu, A.T. Balaban, I. Pascaru, E. Sliam, M. Elian and C.D. Nenitzescu, Tetrahedron, 19 (1963) 1133.
55   K.B. Wiberg and A.S. Fox, J. Amer. Chem. Soc., 85 (1963) 3487.
56   C.F. Cullis and J.W. Ladbury, J. Chem. Soc. London, (1955) 555, 1407, 2850.
57   J.I. Brauman and A.J. Pandell, J. Amer. Chem. Soc., 92 (1970) 329.
58   A.K. Awasthy and J. Roček, J. Amer. Chem. Soc., 91 (1969) 991.
59   J. Halpern and H.B. Tinker, J. Amer. Chem. Soc., 89 (1967) 6427.
60   P.M. Henry, J. Amer. Chem. Soc., 87 (1965) 4423.
61   J. Roček and J.C. Drozd, J. Amer. Chem. Soc., 92 (1970) 6668.
62   K.B. Wiberg and R.D. Geer, J. Amer. Chem. Soc., 88 (1966) 5827.
63   K.B. Wiberg and K.A. Saegebarth, J. Amer. Chem. Soc., 79 (1957) 2822.
64   K.B. Wiberg, C.J. Deutsch and J. Roček, J. Amer. Chem. Soc., 95 (1973) 3034.
65   F. Freeman and N.J. Yamachika, J. Amer. Chem. Soc., 92 (1970) 3730.
66   F. Freeman, P.D. McCart and N.J. Yamachika, J. Amer. Chem. Soc., 92 (1970) 4621.
67a   M. Jáky and L.I. Simándi, J. Chem. Soc. Perkin II, (1972) 1481.
67b   L.I. Simándi and M. Jáky, J. Chem. Soc. Perkin II, (1972) 2326.
68   F.H. Westheimer and A. Novick, J. Chem. Phys., 11 (1943) 506.
69a   F.H. Westheimer and N. Nicolaides, J. Amer. Chem. Soc., 71 (1949) 25.
69b   M. Cohen and F.H. Westheimer, J. Amer. Chem. Soc., 74 (1952) 4387.
70   W. Watanabe and F.H. Westheimer, J. Chem. Phys., 17 (1949) 61.
71a   J. Roček, Collect. Czech. Chem. Commun., 23 (1957) 833.
71b   J. Roček, Collect. Czech. Chem. Commun., 25 (1960) 375.
72   J. Hampton, A. Leo and F.H. Westheimer, J. Amer. Chem. Soc., 78 (1956) 306.

73   M. Rahman and J. Roček, J. Amer. Chem. Soc., 93 (1971) 5462.
74a   J. Roček and A.E. Radkowsky, J. Amer. Chem. Soc., 90 (1968) 2968.
74b   J. Roček and A.E. Radkowsky, J. Amer. Chem. Soc., 95 (1973) 7123.
75   P.M. Nave and W.S. Trahanovsky, J. Amer. Chem. Soc., 92 (1970) 1120.
76   P.M. Nave and W.S. Trahanovsky, J. Amer. Chem. Soc., 90 (1968) 4755.
77   M. Rahman and J. Roček, J. Amer. Chem. Soc., 93 (1971) 5455.
78   K.B. Wiberg and S.K. Mukherjee, J. Amer. Chem. Soc., 96 (1974) 1884.
79   M. Doyle, R.J. Swedo and J. Roček, J. Amer. Chem. Soc., 95 (1973) 8352.
80   F. Hasan and J. Roček, J. Amer. Chem. Soc., 94 (1972) 3181.
81   F. Hasan and J. Roček, J. Amer. Chem. Soc., 94 (1972) 8946.
82   V. Srinivasan and J. Roček, J. Amer. Chem. Soc., 96 (1974) 127.
83   F. Hasan and J. Roček, J. Amer. Chem. Soc., 96 (1974) 534.
84   G. Dyrkacz and J. Roček, J. Amer. Chem. Soc., 95 (1973) 4756.
85   R. Stewart and R. van der Linden, Discuss. Faraday Soc., 29 (1960) 211.
86   K.K. Banerji, J. Chem. Soc. Perkin II, (1973) 435.
87   R. Slack and W.A. Waters, J. Chem. Soc. London, (1949) 594.
88   A.C. Chatterji and S.K. Mukherjee, Z. Phys. Chem. (Leipzig), 210 (1959) 166.
89   J. Roček and F.H. Westheimer, J. Amer. Chem. Soc., 84 (1962) 2241.
90a   G.T.E. Graham and F.H. Westheimer, J. Amer. Chem. Soc., 80 (1958) 3030.
90b   K.B. Wiberg and T. Mill, J. Amer. Chem. Soc., 80 (1958) 3022.
91   A.C. Chatterji and S.K. Mukherjee, J. Amer. Chem. Soc., 80 (1958) 3600.
92   K.B. Wiberg and G. Szeimies, J. Amer. Chem. Soc., 96 (1974) 1889.
93   K.B. Wiberg and P.A. Lepse, J. Amer. Chem. Soc., 86 (1964) 2612.
94   J. Roček and Chiu-Sheung Ng, J. Amer. Chem. Soc., 96 (1974) 1522.
95   F. Freeman, J.B. Brant, N.B. Hester, A.A. Kamego, M.L. Kasner, T.G. McLaughlin and E.W. Paull, J. Org. Chem., 35 (1970) 982.
96   P.A. Best, J.S. Littler and W.A. Waters, J. Chem. Soc. London, (1962) 822.

214

97   J. Roček and A. Riehl, J. Org. Chem., 32 (1967) 3569.
98   J. Roček and A. Riehl, J. Amer. Chem. Soc., 88 (1966) 4749.
99   K.B. Wiberg and R.D. Geer, J. Amer. Chem. Soc., 87 (1965) 5202.
100   K.B. Wiberg and R. Stewart, J. Amer. Chem. Soc., 78 (1956) 1214.
101   R.P. Bell and D.P. Onwood, J. Chem. Soc. B, (1967) 150.
102   J.P. Candlin and J. Halpern, J. Amer. Chem. Soc., 85 (1963) 2518.
103   P. Saffir and H. Taube, J. Amer. Chem. Soc., 82 (1960) 13.
104   F. Hasan and J. Roček, J. Amer. Chem. Soc., 94 (1972) 9073.
105   J.W. Ladbury and C.F. Cullis, Chem. Rev., 58 (1958) 403.
106   S.J. Adler and R.M. Noyes, J. Amer. Chem. Soc., 77 (1955) 2036.
107   G.V. Bakore and S. Narain, J. Chem. Soc. London, (1963) 3419.
108   F.B. Beckwith and W.A. Waters, J. Chem. Soc. B, (1969) 929.
109   D.S. Jha and G.V. Bakore, J. Chem. Soc. B, (1971) 1166.
110   M. Jáky, L.I. Simándi, L. Maros and I. Molnár-Perl, J. Chem. Soc.
      Perkin II, (1973) 1565.
111   L.I. Simándi and M. Jáky, J. Chem. Soc. Perkin II, (1973) 1856.
112   J.C. Sullivan and J.E. French, J. Amer. Chem. Soc., 87 (1965) 5380.
113   E.A. Alexander and F.C. Tompkins, Trans. Faraday Soc., 35
      (1939) 1156.
114   D.H. Rosenblatt, G.T. Davis, L.A. Hull and G.D. Forberg, J. Org.
      Chem., 33 (1968) 1649.
115   M.M. Wei and R. Stewart, J. Amer. Chem. Soc., 88 (1966) 1974.

REACTION INDEX

216